Panzerkampfwagen

GERMAN COMBAT TANKS 1933 - 1945

Chris Ellis and Hilary Doyle

BELLONA

BELLONA PUBLICATIONS
ARGUS BOOKS LTD,
KINGS LANGLEY
HERTS.

First published — 1976

Printed in Great Britain for the publishers, Argus Books Ltd., Station Road, Kings Langley, Herts., England, by The Garden City Press Limited, Letchworth, Herts.

Introduction

THIS book, first in a new Bellona series, is a compact but comprehensive guide to the German tanks developed during the period of the Third Reich, 1933-45. The object is to provide the enthusiast with a model by model record of production vehicles, prototypes, and projects. Chassis numbers and other pertinent information is given wherever possible. Taking advantage of recent new research we have been able to modify some previously published information, making this book more detailed and accurate than any previously published. Some types, such as the PzKpfw 38(t) nA, which have not generally been recorded before are included here for the first time.

The main body of the book presents each series of vehicles in roughly chronological order, but it must be born in mind that much development and production work was, in fact, concurrent. Within the physical limitations of a book of this size we have included as much incidental information as space allows, and personalities and main events are recorded where they affected development. Such aspects of the subject as tactics, and general history cannot be covered, nor can unit organisation in any detail, though these matters have been well recorded in several other books. A brief overall history of German tank development is included for completeness, but this again has been covered in more detail in some other books published in recent years.

To assist the reader in quickly checking on individual models, we have adopted the popular method of introducing each model with a bold heading. Thus to check up on characteristics of, say, the PzKpfw III Ausf M, it is easy enough to scan through the PzKpfw III section and pick out the appropriate heading line.

Specifications of representative models of each production series are included at the back of the book, and multi-view drawings to 1:76 scale are included from the well-known Bellona range. These are by H.L. Doyle and D.P. Dyer.

Acknowledgements

Author and publishers would like to thank the following for assistance in the preparation on this book: W.J. Spielberger, T.L. Jentz C.K. Kliment, D.P. Dyer, James Loop, the Imperial War Museum, London, and Bundesarchiv, Koblenz. The photographs are reproduced either from the authors' personal collections or by courtesy of the Imperial War Museum (IWM) or Bundesarchiv (BA) except where otherwise credited. Negative numbers where known are included with the photo credit.

PzKpfw VI, Tiger Ausf E (BA-4g-8a-27a).

Contents

German Tank Development, 1933-45

IN the 1914-18 war, Germany lagged well behind the Allies in the early development of the tank, and in the development of tank warfare. Their belated efforts to catch up were suddenly stopped by the Armistice, and the limitations imposed by the Versailles Treaty in 1920 (which forbade tanks) officially prevented further development. However, by the late 1920s a lot of enthusiastic theoretical work in tank tactics had been done and dummy tanks were improvised, as detailed in the main text. Vollmer, the leading German tank designer of World War 1 had gone to Sweden, and the Landsverk M-21 was a development of his LK-11 light tank prototype of 1918. Thus there remained some practical contact with tank design and development. Starting in 1926 a series of prototype tanks was built secretly under concealed designations (Leichte traktor and Grosstraktor). The secret Rapallo Treaty between Germany and Soviet Russia in 1926 gave the Germans test facilities for tanks at Kazan in the USSR. In Britain in 1927-28 there was set up the 'Experimental Mechanized Force', virtually the prototype of all modern armoured formations, and leading German theorists in the art of motorised warfare, notably Guderian and Lutz, were influenced by this, and by the writings of the leading British tank theorists of the day, such as Liddell-Hart and Fuller, who had developed ideas in tank warfare from the successful British use of tanks in 1917-18. The British Tank Brigade of the 1931-34 period succeeded the Mechanized Force and further demonstrated how tanks could be used as an armoured mobile force, forming the principal strategic arm of an army in the field.

In 1933, when the National Socialists under Hitler took supreme power, the limitations of the Versailles Treaty were soon disregarded and massive expansion and rearmament of the German forces started. All the ideas formulated since 1920 could now be realised, and the motor transport troops, a corps built up by Guderian and Lutz, formed the nucleus from which the armoured troops (Panzertruppen) were organised. The question of equipping the new armoured forces remained, and took longer to resolve. The prototypes built in the late 1920s and early 1930s were considered unsuitable. While better combat tanks were developed to meet exact requirements, two small training tanks, PzKpfw I and II, were quickly designed and got into service. For the main armament of the panzer divisions two types of tank were developed; a 'light' tank with high velocity gun to fire armour piercing rounds, and a 'medium' tank with low velocity gun to give support fire. To maintain a degree of secrecy these were developed initially under 'camouflage' names to conceal their true functions. The 'light' tank was known as the Zugfuhrerswagen (ZW) or 'comany commander's vehicle', while the 'medium' tank was known as the Bataillonsfuhrerwagen (BW) or 'battalion commander's vehicle.' From 1939 these were known as the PzKpfw III and IV respectively. The BW vehicles were actually in service a little ahead of the ZW, but in 1939-40, in the opening months of World War 2, both types were still in short supply and brunt of the early fighting fell on the PzKpfw I and II, which had been strengthened by the useful acquisition of large numbers of Czech gun tanks of good quality—the PzKpfw 35(t) and PzKpfw 38(t)—obtained when Germany marched into Czechoslovakia early in 1939.

The first three armoured divisions (panzer division, abbreviated to Pz.-Div.) were officially established on October 15, 1935. Guderian had the main influence on organisation and equipment, and it was he who was largely responsible for the doctrine chosen whereby each division was a well-balanced force in its own right, with motorised infantry, reconnaissance troops, assault engineers, and artillery supporting the prime arm, the panzer battalions. Infantry guns, anti-tank guns, and assault engineers were organic at battalion level. Thus a high degree of self-contained mobility was possible, and the addition of a further important element was close-support by air strike just ahead of the attacking panzer division, the Junkers 87 'Stuka' being the main instrument of this. These tactical ideas were far in advance of those of Germany's potential enemies in the late 1930s, notably Britain and France, where divisional organisation was much less integrated. Further experience was gained by Germany in the Spanish Civil War (1936-39) where German equipment and volunteers served with Franco's forces and the so-called 'blitzkrieg' tactics were put into practice. The occupations of Austria and Czechoslovakia in 1938 and 1939 also involved the swift deployment of panzer divisions, though there was no actual combat.

On September 1, 1939, German forces invaded Poland, triggering off the hostilities of World War 2. By this time there were six panzer divisions available plus a Panzer-Lehr (tank demonstration) battalion, committed to battle. The panzer divisions were 1.Pz.-Div., 2.Pz.-Div., 3.Pz.-Div., 4.Pz.-Div., 5.Pz.-Div., and 10.Pz.-Div. On May 10, 1940, these six divisions, which had been transferred to the West, spearheaded the classic blitzkrieg campaign against France and Flanders. Just before this campaign four light (ex-cavalry) divisions were converted to panzer divisions, becoming 6.Pz.-Div., 7.Pz.-Div., 8.Pz.-Div. and 9.Pz.-Div. Of these, the first three were equipped mainly with the ex-Czech tanks and had a reduced complement of tanks. Though still in the minority, many more PzKpfw III and IV were available for this campaign.

Though several projects were being produced in the 1939-41 period which included heavy 'breakthrough' tanks (the DW) and potential successors to the PzKpfw IV (the VK 3001 series), none of these reached the production stage. This was partly due to the struggle to get the PzKpfw III and IV into large scale service, which tied up production facilities in any case without the added complication of introducing further new tyres; but it was mainly due to the satisfaction felt with the existing PzKpfw III and IV and PzKpfw 38(t) which acquitted themselves well in the 1940 campaigns. In fact, all four standard vehicles, PzKpfw I, II, III, and IV, and the Czech PzKpfw 38(t) proved sound and satisfactory in service. In the German-built vehicles, such features as welded armour, Maybach engines and transmission components (except in the Krupp-engined PzKpfw I Ausf A), front sprocket drive, and skeleton-type short pitch cast steel tracks, were common throughout. The Maybach

The severe winter conditions on the Russian Front proved a hard test for men and machines. These are PzKpfw IV, Ausf G on left, Ausf H on right, in winter 1943-44. Note turret skirt armour and late pattern cupola with one-piece hatch on the Ausf H.

water-cooled petrol engines were in-line type for the PzKpfw I Ausf B and II, and V-12 type for the PzKpfw III and IV. They gave a satisfactory average power to weight ratio of around 14 hp/ton. On the two smaller vehicles steering was of the simple clutch-and-brake type, and on the two larger machines steering mechanism was of the epicyclic clutch-and-brake type. There was no 'infantry tank' type such as had been developed by other major powers like France, Britain, and USA, though it is apposite to mention here the assault gun (Sturmgeschutz), initially on the same chassis as the PzKpfw III, which was in service from 1940 in ever increasing numbers. With a limited traverse mount, this was not a tank (and is outside the scope of this book) but it was an important type which was used largely in the infantry support role.

After the 1940 French campaign, there was a reorganisation of the panzer troops to give twice as many panzer divisions. This was done largely by reducing the tank component of each existing division by about half, and adjustments to the supporting arms. Typically there were now two tank battalions and three motorised infantry battalions. However, as the PzKpfw III and IV were now in wide service the theoretical firepower was greatly increased. Two divisions, 5th Light (later 21.Pz.-Div.) and 15.Pz.-Div. were sent to North Africa in 1941 and formed the backbone of the celebrated Afrika Korps. In June 1941 the great invasion of Russia (Operation Barbarossa) began and the hitherto almost unchallenged supremacy of the German panzer divisions began to falter. By now with a 5 cm gun, the PzKpfw III had become the most important German tank at this time and represented over 50% of the strength. However, two vastly superior tanks, the T-34 and KV-1 were in Soviet hands, with 7.62 cm high velocity guns, and the T-34, in

PzKpfw III Ausf M in Tunis, early 1943. Note short 7.5 cm gun and smoke dischargers on turret.

particular took the Germans by surprise even though the first few months of the fighting went in favour of the Germans.

Rapid meesures had to be taken to meet the new situation, and the first direct outcome was the development of a 45 ton design (VK 4501) to produce the famous Tiger tank, with, massive armour and the 8.8 cm gun developed from the Flak 18/36 anti-aircraft gun, which also enjoyed great success in its own right as a formidable anti-tank gun. The Tiger was produced in record time – though work on design and prototypes had started before the invasion of Russia – and the prototype was running in April 1942 with production vehicles in service in the fall of the same year. A superior design, VK 3002, which became the Panther, closely followed the Tiger into production and service. This vehicle was directly influenced by the T-34 and had similar characteristics, notably well sloped armour. The powerful 7.5 cm L/70 gun on the Panther also gave it an edge in firepower over the T-34. By 1943 the qualitative advantage was back in German hands, though in quantitative terms the Germans never again had an advantage in tank strength on any front. Compared to Soviet and other Allied tanks, the Tiger and Panther were superb examples of sophisticated engineering, with such features as overhanging superstructures, interleaved road wheels, and regenerative controlled differential transmission in the case of the Tiger. With massive armour protection, however, they both worked out at more than 10 tons heavier than their originally envisaged weights, and this, plus speed of development, led to numerous 'teething' troubles which rendered them less than satisfactory when first in service.

Meanwhile the existing PzKpfw III and IV were quickly upgunned to meet the threat of the Soviet tanks, and appeared in 1942 with the long 5 cm L/60 gun (PzKpfw III) and long 7.5 cm L/43 gun (PzKpfw IV), while the PzKpfw 38(t) became the basis of various SP gun carriages. The PzKpfw IV Ausf F2 and succeeding models with the long 7.5 cm high velocity gun proved to be the most enduring and widely used of all German tanks for the rest of the war, but the PzKpfw III had reached the limit of its development as a combat tank since it could take a gun no larger than the 5 cm weapon. Aside from specialised roles, the PzKpfw III saw some use as a support tank with low velocity 7.5 cm gun, but virtually disappeared from the panzer troops inventory in late 1943. Another type of vehicle which now appeared in ever increasing numbers was the 'tank destroyer' or 'tank hunter' (Jagdpanzer or Panzerjager) which was evolved from the assault gun by fitting a high velocity gun in place of the low velocity gun of the original assault guns. Several important types of Jagdpanzer appeared on chassis derived from the

Tiger tanks on the Russian Front in 1943. Zimmerit anit-magnetic coating can be seen on nearest vehicle.

PzKpfw III, IV, Panther, Tiger and PzKpfw 38(t) series.

Although tank production was doubled between 1941 and 1943, and peaked in April 1944, the fighting on several different fronts, East and West, imposed a huge attrition rate on the German armoured forces. The Allies built up vast numbers of armoured divisions, and had air superiority over every battlefield from 1942 onwards. Almost unlimited replacements were available for Allied losses, but while German tanks were still skilfully deployed and usually had superiority of firepower on an individual basis, this was insufficient to compensate for the heavy loss rate. Numerous other adverse factors played their part in affecting the fighting quality of the panzer divisions as the war ran to a close, among them fuel shortages, spare parts shortages, disruptions of communications, shortage of raw materials, and bombing of the factories.

The final important tank to see service was the Tiger II, or Tiger Ausf B, an improvement on the original Tiger and with a more powerful 8.8 cm gun. Its huge weight of 70 tons affected its mobility, and it was never available in sufficient quantity to be more than a thorn in the flesh to the Allies. However, the appearance of this tank in 1944 coincided with efforts to produce a superior range of tanks, such as the Panther II and the E series with standardised components and common chasis characteristics. The urgent need for tanks from existing production lines in the rapidly deteriorating war condition, however, precluded the realisation of such plans. An indication of the desperate situation in 1944 is seen in the plan to phase out the PzKpfw IV for the Panzer IV/70 – which was a tank in name only, the tank destroyer type of vehicle being quicker and easier to build in the severely curtailed circumstances of the time.

By 1944 there were 25 panzer divisions under Army command, plus 8 SS-Panzer divisions, and the Hermann Goering parachute corps of two divisions, each with a panzer element. Panzer-Lehr had been upgraded to divisional size in 1943. Army panzer division numbers ran from 1 to 27, but 10 Pz-Div., 22.Pz.-Div., and 27.Pz.-Div. were all wiped out on the Eastern front in 1943, and 18.Pz.-Div. was disbanded. Additionally there was 116.Pz.-Div. and Panzer-Lehr Div. By this time there was great diversity of equipment and organisation, due to the prevailing circumstances, but two tank battalions and four motorised infantry battalions were the most common nucleus of a division.

Popular legend, much disseminated at the time, of the invincibility of the German panzer divisions and their superior numbers in the field is far from the truth. Only in the invasion of Poland in 1939 and the Balkans in 1941 did the Germans have superior numbers. In France in 1940 they were outnumbered by the British and French tanks, and by the British tanks in the Western Desert in 1941-42. In Russia in 1941 they were outnumbered by four to one. Superior tactics and deployment, together with superior firepower (except in Russia in 1941) was responsible for the considerable success the Germans enjoyed in tank fighting throughout the war. Actual German wartime tank production was rather less than that of the British, around a third of that of the United States, and a much smaller fraction still of Soviet tank output. In all the circumstances, therefore, the overall success of the German

The later German tanks were complicated to repair and maintain; hard work is needed to replace damaged road wheels on this mined Tiger.

panzer troops as a fighting arm is even more remarkable.

In retrospect German tanks tended to be over-elaborate and complicated, which absorbed too much time in production and maintenance. In the early days there was no long term provision for mass-production, and production facilities and scale of production was inadequate to meet demand. Design work was extremely thorough, and quality was high. Such failings as there were were largely due to overall policy. Thus there was too little provision for standardisation in the earlier period, and an over-proliferation of types. Both the PzKpfw III and IV were so similar in size that they could profitably have been combined as one vehicle, but by the time this was fully appreciated there was not the development time available and there was only limited progress towards a unified design. At the highest level there was the quest for ever heavier guns and armour which led to money, time, and other valuable resources being diverted to produce vehicles like Maus and E.100 when effort could have been better expended producing a lighter replacement for the Tiger/Panther series.

Procurement authorities

Overall responsiblity for formulating requirements for tank designs, in consultation with the General Staff, was held by the Army Ordnance Office – *Heereswaffenamt*, sometimes referred to as *Waffenamt*. The section of this department specifically responsible for mechanised vehicles, and hence tanks, was Weapons Section 6 – *Wa Pruf 6*. Requirements for new designs were passed either to automotive, engineering, or armaments firms, or to design bureaux such as that of Dr Porsche. Designs accepted and approved were ordered into production through the Reichs Armaments Ministry. There were exceptions to the normal procedure, such as in the case of Maus which appears to have been ordered direct by the Armaments Ministry, Heereswaffenamt having little or no involvement. There was an advisory body, the Panzer Kommission, consisting of service chiefs and leaders of industrial firms involved in tank production. Hitler took a great personal interest in tank design and often intervened in the deliberations, or sent orders or suggestions. To be fair, some of his ideas were useful or ahead of his advisers, notably on the question of more powerful guns and heavier armour. On the other hand his interference sometimes delayed or obscured other important matters, and antagonised the staff.

Tank nomenclature

Prior to 1939, most new tanks were known under 'camouflage' names to conceal their true purpose. Thus the PzKpfw I was originally known as the *Landwirtschaftlicher Schlepper* – agricultural tractor – *La S* in abbreviated form. Further models were indicated by a number – thus *1A/La S, 1B/La S,* and so on. The PzKpfw IV was similarly known as a *Bataillonsfuhrerwagen* – battalion commander's vehicle – the production models being *1/BW, 2/BW*, and so on. From September 1939 all existing service vehicles were redesignated as *Panzerkampfwagen* – armoured fighting vehicle or tank – with a model number. This was usually abbreviated to *Pz.Kpfw, PzKpfw., PzKw,* or *Pz*, the first style being used in this book. Thus the *BW* vehicle became the *PzKpfw IV*. Models developed within the production period were indicated by letters. The word for 'model' was *Ausfuhrung*, usually abbreviated to *Ausf.* Thus: *PzKpfw IV Ausf B, PzKpfw VI Ausf E,* and so on. Vehicles standardised and accepted into service also had an ordnance inventory number allocated by the Ordnance Dept. This number was the *Sonderkraftfahrzeug* – special vehicle number – normally abbreviated to *Sd Kfz*. A typical example: *PzKpfw VI Ausf E(Sd Kfz 181).* Certain vehicles were allocated type names as well, thus: *Tiger, Panther.* An example: *PzKpfw VI Tiger Ausf E (Sd Kfz 181).*

When a development requirement was formulated by Heereswaffenamt, it was allocated a code in the *VK* series. *VK* was an abbreviation for *Vollkettenkraftfahrzeug* – fully tracked experimental vehicle. The code number indicated its weight class, the number of the prototype requirement, and the contractor or designer. Thus *VK.3001(H)* indicates 'fully tracked experimental vehicle in the thirty ton class, first prototype, designed by Henschel'. A further prototype in the same weight class would have been designated *VK.3002(H)* and so on. Makers usually had their own designations for the same vehicle, though this was not an official military designation. Thus *Typ 100 Leopard* was the Porsche name for the prototype which the Heereswaffenamt knew as *VK.3001(P).*

In this book, where the same vehicle was known either simultaneously or at different times by different designations, we give the various designations in the first (bold type) heading, with the different designations separated by an oblique stroke (/). Thus: **Typ 9/ZW / PzKpfw III Ausf L (Sd Kfz 141/1).**

Technical Characteristics

WHILE it is not physically possible in a book of this size to give an extensive coverage of the technical and physical aspects of all German tanks of the World War 2 period, in this section we outline in simple terms some of the more important features.

Main armament and Ammunition

Though obviously differing in size and detail, the mechanisms of all the main German tank guns were similar. They are semi-automatic in operation and were fired electrically, though there was provision for firing by percussion. A typical breech layout (for the 5 cm gun) is illustrated.

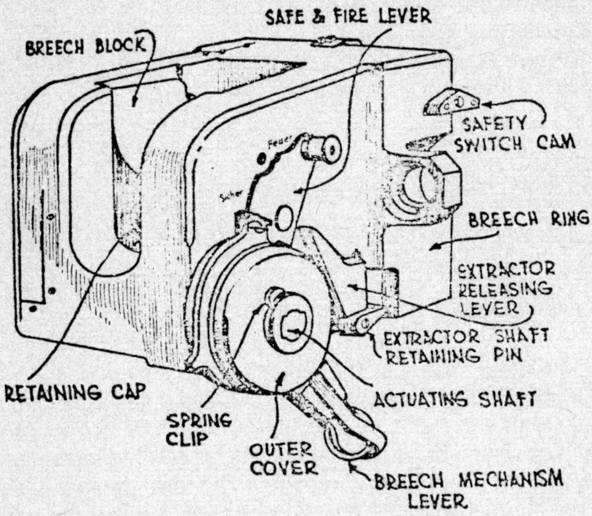

2 cm KwK 38: This was the main armament of vehicles in the PzKpfw II series. Unlike larger tank guns it was loaded from the left side by magazine, the flat curved magazine holding ten rounds. It was derived from the Flak 38 anti-aircraft gun. This gun was fired by a foot pedal. Single shot or automatic fire could be selected. It was reloaded simply by changing the magazine.

Ammunition

HE tracer; yellow projectile, black annulus, round weight 4.2 oz.
AP tracer: black projectile, yellow band, black annulus, round weight 5.2 oz.
AP 40: unpainted aluminium body with pointed nose, red band above driving band, black varnish below, round weight 3.6 oz.
(HE: high explosive, AP: armour piercing).

5 cm KwK L/60: This was the 'long' 5 cm weapon fitted on later models of the PzKpfw III. It was mounted co-axially with a MG 34 which was belt-fed and fired by a foot pedal by the gunner who was on the left side of the gun (the MG 34 being to the right of the gun). A telescopic sight TZF 5e was used with this mounting and was provided with three range scales which could be selected to suit either HE ammunition, AP, or the MG 34.

Ammunition

APC shell: black, marked 'Pzgr. Patr.39'.
AP 40 shot: black, marked 'Pzgr. Patr.40'.
HE shell: olive green, marked 'Sprgr. Patr.38'.
(APC: armour piercing, capped).

7.5 cm KwK L/24: This was the short low velocity gun fitted to early models of the PzKpfw IV and the final models of the PzKpfw III. It had a co-axial MG 34 and sighting telescope (TZF 5b) arranged as described for the 5 cm gun.

Ammunition

APCBC shell: black, marked '7.5 cm. Pzgr.'.
Hollow charge shell: white or green, marked '7.5 cm. Gr. 38'.
HE shell: olive green, marked '7.5 cm. Gr. 34'.
Smoke shell: olive green, marked '7.5 cm... Nbgr.'.
(APCBC: armour piercing capped. ballistic capped).

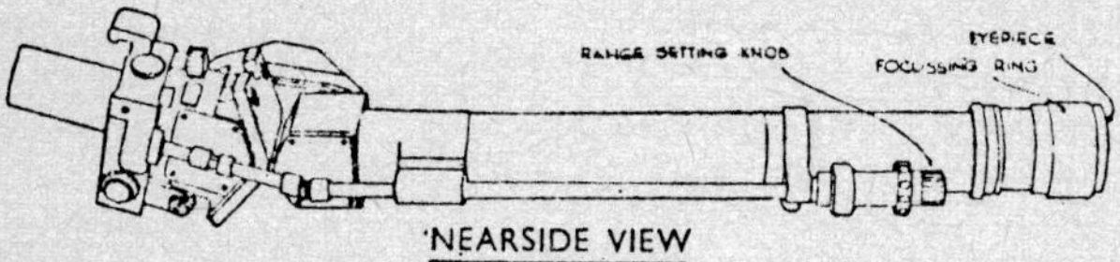

Typical sighting telescope (TZF 5b).

7.5 cm KwK 40 L/43: This was the 1942 'long' high velocity 7.5 cm gun fitted to the PzKpfw IV from the Ausf F2 onwards. Very early guns had a single baffle muzzle brake, but a double baffle muzzle brake was standard. The arrangement for the co-axial MG 34 and the telescopic sight was similar to that described for the 5 cm gun, except that the three range scales on the TZF 5f sight were marked for APCBC, HE, and AP 40; it was necessary to estimate range for the MG 34 and for firing smoke, though the estimation could use the HE scale as a basis.

Ammunition

APCBC shell: black, marked 'Pzgr. 39'.
AP 40 shot: black, marked 'Pzgr. 40'.
Hollow charge shell: olive green, marked 'Gr.38'.
HE shell: olive green, marked 'Sprgr.34'.
Smoke shell: olive green, marked 'Nb'.

8.8 cm KwK 36 L/56: This was the gun carried in the Tiger Ausf E. It was derived from the Flak 18 and 36 anti-aircraft guns, but embodied some detail changes, including a change over from percussion to electrical firing. A co-axial MG 34 was carried in the same mount. The sight used was of binocular type (TZF 9b). Range scales for APCBC and the MG 34 were incorporated on graticules within the sight, but the range for HE was put on a clinometer on the gun cradle and laid by elevation against the range scale. A rangefinder and scissors telescope were carried in the tank and could be used for more accurate rangefinding when sustained HE fire was necessary.

Ammunition

APCBC shell: black with white tip, marked 'Pzgr. 39'.
AP 40 shot: black, marked 'Pzgr. 40'.
HE shell: yellow, marked 'Spgr'.
Hollow charge shell: olive green marked 'Gr.39 H/I'.

Machine guns and secondary armament

The standard secondary armament was the MG 34 machine gun, which in various forms and mounts was used by all arms of the Wehrmacht. For tank use it was fitted in a co-axial mount with the main armament and (except in the PzKpfw I and II) it was also mounted to the right side of the hull front and was the responsibility of the radio operator. The hull machine gun was,

MG 34 in Kugelblende mount, hull front (Tiger).

with some exceptions (eg, PzKpfw V Ausf D) carried in an armoured ball mount (Kugelblende) which incorporated a sighting aperture. The form of the ball mount varied with the tank. In early versions the protruding end of the barrel was fully exposed, but in later versions of the PzKpfw III and IV it was fitted with an armoured sleeve. Ammunition could be belt-fed (usual with the co-axial gun) or drum fed (usual with the hull gun). From 1943 onwards there was a third mount on most tanks, Fliegerbeschussgerat, which

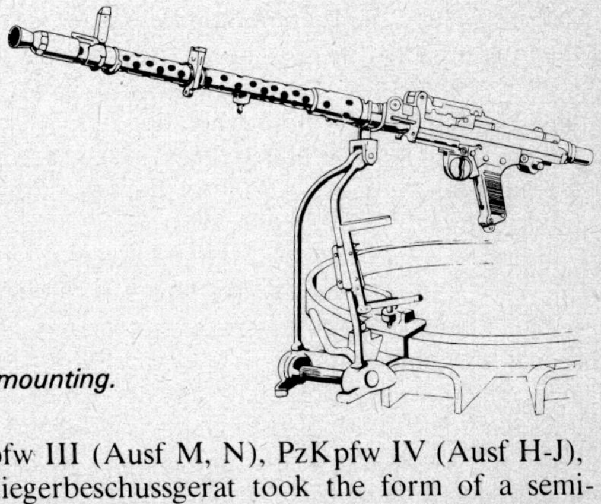

MG 34 on Fliegerbeschussgerat mounting.

was fitted to late model PzKpfw III (Ausf M, N), PzKpfw IV (Ausf H-J), Panther, and Tiger series. Fliegerbeschussgerat took the form of a semicircular rail which was fitted round the forward arc of the commander's cupola. An adjustable mounting bracket could be traversed round the rail, or could be clipped in position at any point on it. The mounting allowed a traverse of 60 degrees, and an elevation of 80 degrees. It was developed in 1942 to give tanks a degree of self-protection from air attack. The commander operated a MG 34 attached to the mount, exposing only his head and shoulders. The drum magazine was usually employed with the MG 34. Gun and mounting brackets were stowed in the tank when out of use.

A further important weapon, mainly fitted to tank destroyers and assault guns,which also equipped the Panther and Tiger models was the Nah-

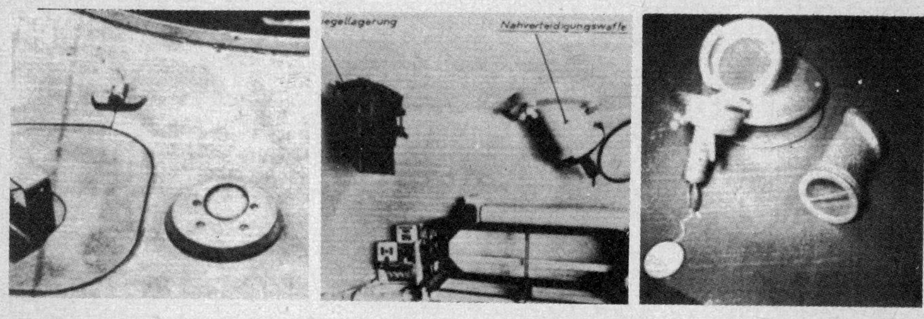

Nahverteidgungswaffe fitted to Panzer IV/70 (left), inside the turret roof of a Panther (centre) and dismantled to component parts (right).

verteidgungswaffe (close-in defence weapon) which was a grenade projector fitted inside the turret roof. The short projector barrel was angled at about 60 degrees to the turret roof and could traverse through 360 degrees. It was hand operated, and fired by percussion via a simple cocking ring. It fired a 92 mm calibre smoke generator (Nebelkerzen) or a high explosive charge; the former was used for concealment, the latter for close defence against tank hunting infantry. It replaced external smoke projectors fitted on earlier tanks and had the advantage of being loaded and fired from inside the tank.

Prior to introduction of the close defence weapon, the standard smoke discharger was the 90 mm NbKWg which was fitted in three-barrel groups to the turret sides. This weapon fired smoke candles only; it was fired electrically from inside the tank and had to be re-loaded from the outside. The 90 mm NbKWg was introduced on the late PzKpfw III and IV, the Panther Ausf D and Tiger Ausf E. Before this, rear smoke emitters were a standard fitting for all tanks, and description of the installation is given in the section of this book covering the PzKpfw III.

German tank wireless

The standard German tank radio equipment was known as *Fu 5 (Funk:* radio) and consisted of a 10 watt transmitter, an ultra short wave receiver, and related accessories. All tanks carried Fu 5, and a second set, Fu 2, which consisted essentially of the receiver set of the Fu 5, was found in the PzKpfw I and some other tanks. Command and observation vehicles carried a variety of other sets, depending on their function, as outlined for specific vehicles in the main text.

Fu 5 was designed for short range communication inside tank regiments, using RT and MCW transmission. It operated on a frequency band of 27,200-33,300 Kc/s in a single range. Power came from a 12 volt battery with a separate transformer for the receiver and transmitter. The approximate range was 2½ miles when stationary and 1½ miles on the move, the RT. For MCW these distances were 3½ miles and 2½ miles. A 2 metre long rod aerial was carried by all tanks fitted with Fu 5.

In most types of tank the radio set was used also in conjunction with an intercom telephone *(Bord),* which provided the commander, radio operator, and driver with internal and external voice contact facilities. The same microphone and headset was used both for the intercom and the Fu 5. There was a special switch on the transmitter marked 'Bord und Funk' which enabled both the commander and the radio operator to transmit messages and the driver to hear these messages. Internal communication was possible at the same time though all conversation was transmitted with the set in this mode. The same intercom facility could be used with Fu 2 except, of course, that there was no transmission facility.

Mechanical components

The PzKpfw I Ausf A had a Krupp air-cooled 'flat four' engine, but from the PzKpfw I Ausf B onwards all German production tanks had a water cooled petrol engine specially made for tanks by Maybach. On the PzKpfw IB and II, this was an in-line engine, but from the PzKpfw III on a more powerful V-12 type was used, and this was progressively uprated to suit the increasing weight of later types like the Tiger and Panther. Cylinder blocks and crank case were an aluminium casting and the cylinder head was cast iron. All

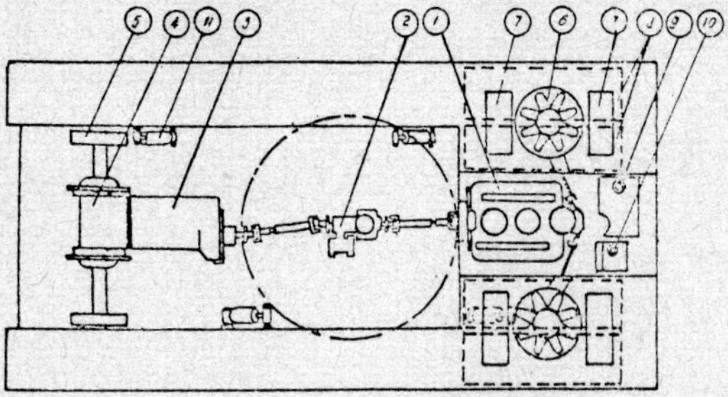

1. Engine
2. Turret base junction.
3. Gear box
4. Steering gear.
5. Steering brake.
6. Fan.
7. Radiators.
8. Fuel tanks.
9. Fuel filler cap.
10. Radiator filler cap.
11. Hydraulic shock absorber.

Typical mechanical layout of a German tank, in this case the Panther.

tanks were fitted with an inertia starter and an electric starter. The engine was designed to use low grade (74 octane) petrol. Large radiators and fans were fitted in the engine compartments adjacent to the engines, Prominent air filters were a necessary and noticeable engine feature. The Ordnance Dept remained faithful to the Maybach engine through to the war's end, though

ABOVE: Fu 5 tank radio. BELOW AND LEFT: Maybach HL 230 P30 V-12 tank engine.

various attempts were made by automotive makers, such as Daimler-Benz, to get a diesel engine adopted as standard. The early tanks, PzKpfw I and II, had the very simple type of clutch-and-brake transmission but more sophisticated systems of the ZF and Variorex (automatic) type were used subsequently, culminating in the very fine Maybach-Olvar fully regenerative transmission in the Tiger which was also planned for the Panther II.

Czech tanks were an exception, and even in variants produced until 1945, the original reliable 5 cylinder in-line petrol Praga EPA engine was used.

Camouflage and markings

From 1943 a special anti-magnetic paste (a cement-like substance) was applied to all main external surfaces of tanks to prevent the affixing of anti-tank mines by infantry tank hunter groups. The method of application, with a toothed spatula, gave the characteristic rough surface visible on vehicles so treated. This material was known as *Zimmerit.* It was sometimes left in its unpainted form (cement colour) but was more usually painted over in the base finish of the tank.

Newly manufactured vehicles were given a base coating of red oxide primer, which was rust coloured, during manufacture. Factory finished vehicles were normally sprayed in a single colour. Any additional colours for camouflage purposes were applied by the units. The following colours were standardised:

Earth Brown RAL 8020; Field Grey RAL 7008; Dark Grey RAL 7027.

From 1935 - 40 vehicles could be seen either in one of these single colours, or RAL 8020 combined with one of the other two in a disruptive pattern. In 1941 application of a fourth basic colour started for vehicles sent to North Africa; this was Sand Brown RAL 8000. This was used either on its own or with a mottle application applied from one of the other colours.

From February 18, 1943, a new system came into effect whereby the previous colours were replaced by a single overall colour of Middle Sand (Mittelsand) applied at the factory. A kit of water based soluble paints in paste form was issued with each tank consisting of the following colours: Olive Green, Red Brown, Dark Yellow, and White, 2 kg of each. This was mixed with water or petrol and applied by spray gun or brush to suit operational conditions. White water based paint had been issued from 1941 as a camouflage for cold weather theatres, such as the Russian front. It was possible to vary the colour scheme from day to day if required. Use was also made of captured Allied paint stocks in theatres where these were available.

From mid 1944 the old dark colours were re-introduced for heavy vehicles (ie. Tiger, Panther) due to the need to conceal these units from the air as much as possible, since they had become prime targets for Allied rocket-armed fighters, specially on the Western front.

The chassis number was frequently stencilled on the right front of the vehicle when new, but this was often subsequently obscured by camouflage painting. A small panel was stencilled on the superstructure side lettered:

Gef.Gew(Number) (= fighting weight)

Ve.Kl.:(Number) (= load class)

These were for rail shipping and bridging information mainly. Panther and

ABOVE: Callsign R01 identifies a regimental commander's BefPzWg Panther Ausf G. Typical camouflage application shows both dark colours and mottle over basic Mittelsand. National insignia is carried on superstructure side.

LEFT: Runic divisional sign (of 4.Pz.-Div,) on superstructure front, and chassis number on glacis plate are in typical positions on this PzKpfw III Ausf G. This is a vehicle originally waterproofed for Operation Sealion, and frames for watertight covers can be seen round the mantlet and Kegelblende.

Tiger carried only the load class information. Again, though factory applied, this panel was frequently overpainted or obscured.

National insignia was a plain white cross applied for the Polish campaign of 1939. This showed to the rear, front, sides, and top. Thereafter a black cross with white outline was used, similar to that used on aircrafts. Sometimes, particularly later in the war, only a white outline type of cross was used. The cross normally showed to the rear and sides only and was painted on the hull or superstructure. There were exceptions, such as application to the turret on some Tiger Ausf B; however, turret application was normally reserved for marking captured vehicles in German service. For air recognition a German national flag was normally used on turret or superstructure top, but local use was also made of coloured cloth panels to suit operational requirements.

The first three panzer divisions adopted 'traditional' formation signs related to their origin; for example 3.Pz.-Div. had the brown bear of Berlin. From 1939, however, formalised runic formation signs were ordered for all panzer divisions and these were painted in white, yellow, or black (depending on background colour) on the front and back of each vehicle (usually left front and right rear but much depended on the vehicle or unit). In 1941 there was a change of allocation of these formation signs. It is not possible to show all signs here, but some examples are given. In some cases the senior divisions also displayed their pre-1939 formation signs or other non-runic signs. Some

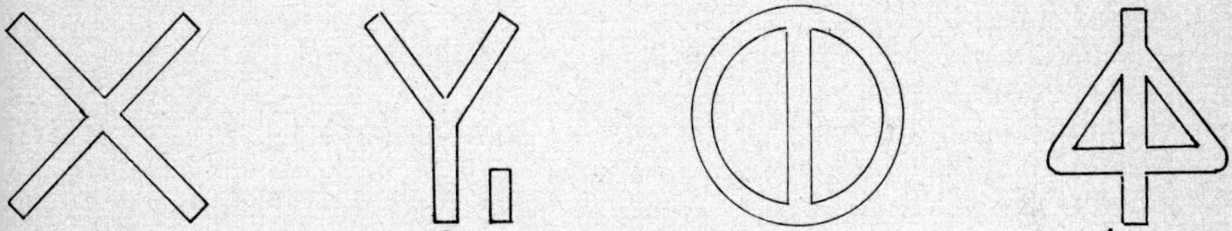

1 2 3 4

Typical runic divisional signs, in use from late 1940. (1) 5.Pz.-Div. (2) 8.Pz.-Div. (3) 11.Pz.-Div. (4) 15.Pz.-Div.

vehicles or units also carried individual markings, notably company markings, but the form of these greatly varied. Some tanks indicated 'kills' by white bands painted round the barrel.

The tactical sign of a tank was its turret number. Tanks did not normally carry the tactical signs used on non-armoured vehicles (though there were exceptions). Until 1940, the vehicles tactical number was painted on a detachable rhomboid-shaped plate (itself the tactical symbol for a tank) and this was clipped to the vehicle to show to the sides and rear. Thereafter the number was more usually painted in large digits on the turret sides, either in 'solid' or white-outlined form. The large numbers were adopted as a result of war experience and enabled a unit commander to quickly call up one of his vehicles by radio and pick it out more easily across the battlefield. The number was painted to show to the sides and rear. The three digits signified the *Kompanie* (first digit), the *Zug* (platoon) (second digit), and individual vehicle in the platoon. Thus 232 indicated, second company, third platoon, second vehicle in the platoon. When there was a tenth company with tanks, there would be a four digit number accordingly. For example the tanks of a panzer-grenadier division were in the tenth company. Thus 1012 would indicate tenth company, first platoon, second tank in platoon.

Command tanks (which included staff tanks, eg, adjutant) were marked with R for the regimental command staff. ROl would be the regimental commander, followed by RO2, RO3, etc, allocated to other command staff officers. Battalion staff vehicles were indicated by the Roman numeral I for the first battalion, and II for the second battalion (*abteilung*) within a regiment. Examples I01, I02, II01, II02, etc.

Sometimes only one or two digits were carried, particularly where companies were operating independently and these indicated simply the platoon and individual vehicle (two digits) or an individual vehicle (one digit).

Manufacturers

The principal producers of tanks and the associated components are listed below. In the case of the important tank production companies we give the full names and locations of the firms concerned, preceded by the 'short' name for the company (where applicable) as is most commonly used in text references. The types produced by the respective tank production firms are also listed.

Tank Production Firms

Alkett:Altmarkische Ketten-Fabrik Gmbh, Berlin	PzKpfw I,II
BMM:Bohmisch-Marische Maschinenfabriken AG, Prague	PzKpfw 38(t)
Daimler-Benz:Daimler-Benz AG, Berlin	PzKpfw I,II,III, Panther
FAMO/Famo:Fahrzeug und Moterenwerke, Breslau, and Famo-	PzKpfw II,III
Ursus, Warsaw	PzKpfw II
Henschel: Henschel & Sohn Gmbh, Kassel	PzKpfw II,III,III,Tiger Panther, Tiger Ausf B
Kraus-Maffei, Munich	VK 1601, VK 601
MAN: Maschinenfabrik Augsburg-Nurnburg AG, Nuremburg	PzKpfw I,II,III,Panther
MIAG:Muhlenbau-Industrie AG,Braunschweig	PzKpfw II,III,Panther
MNH:Maschinenfabrik Niedersacshe,Hannover	PzKpfw III,Panther
Ni-werke/Nibelungenwerke: Steyr-Daimler-Puch, St Valentin, Austria	PzKpfw IV,Tiger(P)
Skoda:Skoda-Werke,Pizen & Prague,Czechoslovakia	PzKpfw 35(t)
Vomag, Vomag-Maschinenfabrik AG, Plauen	PzKpfw IV
Wegmann; Wegmann Wagenfabrik,Kassel	PzKpfw I,II,III

Major Components Suppliers

(Engines:)
Maybach, Berlin
Daimler-Benz, Berlin
Auto-Union,Siegmar

(Gearboxes:)
Adler, Frankfurt-Main
ZF,Berlin

(Weapons:)
Rheinmettal-Borsig,Dusseldorf
Franz-Garny,Frankfurt-Main
Karges-Hammer,Brandenburg
Krupp,Essen
Skoda,Plisen

Development Firms

Adler, Frankfurt-Main
Krupp, Essen
Porsche, Stuttgart
Rheinmetall, Dusseldorf

(Armour Plate:)
Bismarkhutte,Oberschlesien
Bohler,Kappenburg
Brandenburgischen Eisenwerk
Deutsche Edelstahl,Hannover
Dortmund-Hoden Huttenverein
Eisenwerk Oberdonau,Linz
Harkort-Eiken,Hagen
Krupp, Essen
Poldihutte,Komotau
Ruhrstahl,Hattingen
Witkowitz Bergbau und Eisenhutten

1. Early Developments

UNDER the terms of the Versailles Treaty of 1919, which came into force in January 1920, harsh limitations were placed on the size of the German Armed Forces (Reichswehr), consistent with Germany's status as a nation defeated in war. The Army was limited to 100,000 men with no tanks of any kind. However, even in the early years of peace, the Army showed determination and zeal in bringing itself to a peak of efficiency and making up for the restrictions which the victorious Allies had imposed. The commander-in-chief of the Army (Chef der Heeresleitung), General Von Seeckt, who held this post until 1926 was a man of some vision and imagination; he realised that mobility and flexible response would probably be the key to success in any future European conflict. Von Seeckt drove home to his junior commanders his ideas on mobility, and held realistic exercies which kept the troops keen and interested—this at a time when the Allied armies were being run down and small regular forces had reverted to leisurely peace-time garrison and colonial duties.

Armoured cars were allowed for policing and frontier patrol work and the bulk of the infantry was organised into motor battalions—seven battalions each with 15 armoured cars—when the Army was officially re-constituted during 1920. These armoured cars were more specifically armoured troop carriers and carried 12 men each in addition to the crew. In 1926 a secret treaty was signed between Germany and the Soviet Union, the Rappallo Agreement, whereby the Soviets gave testing facilities to the Germans at a trails centre at Kazan on the River Volga. Here, well away from the prying eyes of the Inter-Allied Control Commission, some potential designs for armoured vehicles could be tested. Rheinmetall, Krupp, and Daimler-Benz were given contracts to the Army Ordnance Office (Heereswaffenamt) to produce two prototypes each of a large tank which was known under a 'camouflage' name of 'Grosstraktor' (large tractor). A further contract was for a light vehicle, known as 'Leichtertraktor' (light tractor). Captain Pirner of the Army Ordnance Office was in charge of procurement. The Grosstraktor weighed 20 tons, had a 7.5 cm gun, a speed of 12 mph, machine guns for all-round defence and enfilading fire, and a propulsive capability in water, the vehicle itself being able to float. The 'light weight' vehicle was to weight 9 tons and have a 3.7 cm gun. Contracts for the prototypes were placed in March 1927. Concurrently specifications were put out for some sophisticated armoured cars, classed at the time as 'wheeled tanks', and the 8 and 10 wheel Daimler-Benz and Bussing-NAG prototypes resulted. Little technical record of these early tank prototypes has survived but the designs are summarised below.

Grosstraktor I: This vehicle had a 10.5 cm gun as main armament, presumably to test its suitability compared with the 7.5 cm gun. The hull was of the rhomboid shape evolved in the successful British heavy tanks of World War 1. An aircraft engine of 300hp was fitted. The main armament was mounted in a fully traversing turret set well forward on the hull, the engine was set just aft of the vehicle, and there was a rear compartment with an auxiliary machine gun turret, mounted so as to enfilade enemy trenches as the tank crossed over them. A passageway on the right side of the vehicle connected the forward compartment to the rear compartment. The vehicle was amphibious (or was intended to be) having a bouyancy capability from its own volume. Water propulsion was by a propellor below the rear hull which was clutched into the final drive, as required, by the rear turret gunner. Whether or not the amphibious capability was over tested is not known, but the requirement was dropped, though propulsive units were constructed. Ferdinand Porsche was the designer. Built by Daimler-Benz.

Grosstraktor II: This was built by Rheinmetall-Borsig and is, perhaps, the best-known of the Grosstraktor series, having been the most widely illustrated. A short low velocity 7.5 cm gun was fitted as main armament, and the machine gun turret was at the rear. It is known that this particular model was fitted with American Cletrac steering. A 250 hp engine gave a speed of 40 kph.

Grosstraktor III: The Krupp vehicle in the series, this was again armed with a 7.5 cm gun. It was similar to the other vehicles but had noticeably larger mud chutes than the others.

All the vehicles were built in mild steel, though 13 mm armour plate was specified for full production standards. MG 08 guns were fitted, one in the rear auxiliary turret, one in the main turret, and provision for a third in the hull front. Assembly of all six prototypes—two from each firm—was done at the Rheinmetall factory in Unterluess, for security reasons though the components were made in the respective 'parent' factories. All were fitted with spring suspension. The prototypes were all delivered in the summer of 1928, but after trials there was no immediate production order. This was probably connected with the great economic slump of the 1929-30 period which led to a severe cut-back in armament production, even on the limited scale contemplated in Germany.

Leichtertractor: To satisfy the 'light weight' requirement, three prototypes each of the 'Leichtertraktor' were orderd from Rheinmetall and Krupp in 1928. The requirement initially was for two of the prototypes from each firm to be completed as gun tanks with turrets, while the third was to be completed without a turret for use either as a munition carrier or weapons carrier. Rheinmetall's prototype appears to mave been the most successful contender, and trials were conducted in 1930. The vehicle had a crew of four, weighed 8.5 tons, had a 3.7 cm L/45 gun, and was to have 13 mm armour. The engine was set towards the front of the vehicle with the turret and fighting compartment to the rear, a similar dispostion to the LK II light tank which was in prototype stage in 1918 and had been built in Sweden in the 1920s. The running units and hull shape were similar to the Rheinmetall Grosstraktor II. Speed was about 12 mph. Like the Grosstraktor, the Leichtertraktor was a well-engineered vehicle, was fully gas-proof and had a good cross-country performance. There was an initial plan to build a series of 17 pre-production

Grosstraktor I (Daimler-Benz); rear view showing auxiliary MG turret (W. J. Spielberger).

Grosstraktor II (Rheinmetall).

Grosstraktor III (Krupp).

Leichtertraktor.

vehicles to equip one company for full troop trials, and a long term plan called for 289 vehicles of this type. However, none of these aims came to fruition.

Both the Leichtertraktor and Grosstraktor designs were, by 1932, considered to be unsuitable, being too slow, and with the commander seated in the hull rather than the turret, where his visibility and control were restricted. This did not accord with the latest tactical requirements, and new plans were put into effect by the new Inspector of Transport Troops, General Lutz and his chief of staff, Colonel Heinz Guderian.

Dummy Tanks

While all these efforts were taking place, the Army was not idle in familiarising itself with tank tactics. The powerful influence was Colonel Heinz Guderian, for some years a staff officer of motorized troops. From 1922 onwards he made a detailed study of the writings of the various early proponents of armoured warfare, notably Liddell Hart and Fuller in England, and he analysed the tank operations carried out by the Allies on the Western Front in 1917-1918. Colonel Guderian had thought of using dummy tanks in the summer exercises of 1927 with troops pushing round canvas mock-ups. Then he had some field cars disguised to resemble tanks by adding outlines of board and sheet metal. The idea was successful and at least one complete dummy 'tank company' was thus equiped for the 1928 summer wargames. NSU 'Dixi' and Adler cars of the Kubelwagen type were the main vehicles used. Thereafter the use of dummy tanks as a training aid became accepted—even after real tanks were available in quantity—since there was saving in money and manpower. A dummy tank for training purposes had become a standard type of vehicle by 1940 with its own place in the ordnance inventory. In 1944-45 when equipment was again in short supply, some of these dummy tanks were actually deployed in action against the Allies on a few occasions.

Heavy Tanks

The experience with the Grosstraktor designs was clearly used in designing some new heavy tanks in 1933 when the decision to equip full scale armoured divisions had been taken. A whole family of tank types was envisaged and those that went into production are outlined on subsequent pages. The 'heavy' class of vehicle in this new generation of types did not, however, proceed beyond the prototype stage. Designated at first as 'Neubaufahrzeuge' (NbFz) ('new construction vehicle'), some five prototypes were built by Rheinmetall and Krupp.

Neubaufahrzeuge/Panzerkampfwagen VI: The first two vehicles of the five NbFz prototypes had a 7.5 cm gun mounted co-axially above a 3.7 cm gun and the turret was of a distinctive cast shape, made by Rheinmetall, the builders. A rail-type radio aerial was affixed to the turret sides. These initial vehicles were made of mild steel, up to 14.5 mm thick. In 1939 the 'PzKpfw VI' designation was given to this vehicle for propaganda purposes, as these prototypes were much photographed referred to as the 'new heavy tank.'

TOP TO BOTTOM: Push-along dummy tanks, 1927 exercises. First motorised dummy tanks, 1928 exercises. Early standard dummy tanks on NSU 'Dixi', 1931 exercises.

Neubaufahrzeuge/Panzerkampfwagen V: Three of the five prototypes were of this type. Armour plate was used and the turrets were built by Krupp. The armament in this model was a 7.5 cm gun co-axially mounted with a 3.7 cm gun, which overcame difficulties experienced with the original mounting. There were two auxiliary turrets similar to the PzKpfw I turret, one at the front to the right and one behind the main turret and to the left MG 13 machine guns were mounted in pairs in these small turrets. The propaganda designation 'PzKpfw V' was promulgated in 1939. These tanks served in Norway in 1940 and Russia in 1941. The armour plate was up to 20 mm thick.

The physical characteristics of the NbFz were clearly influenced by contemporary Soviet (T-28) and British (Independent and Vickers 16 tonner) designs, following the style and layout of these multi-turreted vehicles quite closely. The engine was a 360 hp 6 cylinder Maybach unit with drive to rear sprockets and there was a six-speed gearbox. Some vehicles had Wilson epicyclic gearboxes, the others Cletrac. The hull construction was riveted, the suspension as for Grosstraktors, with bogies in pairs, suspended on short sprung beams, five pairs each side. The side sponsons included access/escape doors, and there were also escape hatches in the superstructure sides and all turrets. The sponson sides supported four return rollers. A crew of seven was carried—commander, driver, two machine gunners, two gunners, and a radio operator. The weight was about 24 tons, the weight limit being 25 tons for AFVs in Germany at that time in order to meet bridging limitations.

By the time the first two prototypes were completed in 1934 a programme of construction for the new armoured divisions had already been worked out, and more modern designs had been ordered (ie. PzKpfw III and IV). However, during 1939-40 the vehicles were well published for propaganda purposes leading Germany's enemies to suppose that the vehicles were in being in service strength. The vehicles formed the centre-piece of the Wehrmacht stand at the Berlin Motor Show in February 1939, and it appears to have been here that the PzKpfw V and VI designations were first publicised. In April 1940 the three armoured vehicles were shipped to Oslo and pictures of them symbolising the armed power of Germany were published in newspapers round the world, though the vehicles in the pictures were the only ones of their type in existence. After the campaign in Norway the two surviving vehicles were eventually returned to Germany and documentary evidence shows that they were ordered to be scrapped some time in 1941. However, battle reports indicate that they were actually destroyed in the first weeks of fighting in Russia in 1941.

OPPOSITE, TOP TO BOTTOM: NbFz/PzKpfw V in service, Norway 1940. NbFz/PzKpfw VI with the mild steel turret and vertical co-axial mount. NbFz/PzKpfw V on test in 1939 with no MGs in auxiliary turrets.

2: Panzerkampfwagen I

By late 1931, Guderian became Chief-of-Staff to General Lutz, Inspector General of Motorized Troops, and these two officers were thus in a good position to influence further policy and developments. Armoured Divisions (Panzer Divisions) as envisaged by Guderian were based on the British model which had emanated in the Experimental Mechanised Force of 1927 (it was later called the Armoured Force), the prototype of all modern armoured formations. In the early part of 1931 Guderian was given the chance by Lutz to organise one of the Army's motorized battalions (3rd Prussian) on the lines of a 'model' Panzer Brigade. No 1 Company was equipped with armoured cars and No 4 Company had motor-cycles, the two together depicting an armoured reconnaissance battalion for exercise purposes; No 2 Company had dummy tanks, representing a tank battalion, and No 3 Company had dummy wooden guns representing an anti-tank battalion. With this formation Guderian was able to demonstrate and evaluate armoured force tactics, though with a few notable exceptions many senior officers were sceptical of Guderian's claims. In his new staff position later that year, however, Guderian was better placed to overcome opposition; cavalry-minded senior officers made an attempt to retain the importance of cavalry, but eventually Guderian was able to push through the case for Panzer Divisions, with a longer term plan for a complete Panzer Corps.

None of the Grosstraktor or Leichtertraktor prototypes accorded with the ideas of Guderian or Lutz – the position of the commander inside the hull, as in World War 1 tanks, was a particular tactical disadvantage. Two new types of tank were therefore planned, one a light tank with armour piercing gun to equip the three light companies of a tank battalion, undertake reconnaissance and spearhead the attack, and one medium tank (about 20 tons) with a 7.5 cm gun to equip the medium company of the battalion. The function of this tank was to support the lighter tanks by engaging targets out of range of the small calibre guns of the lighter tanks.

For the light tank Guderian wanted a 5 cm gun with armour-piercing capacity, but a compromise was made on a 3.7 cm gun since the infantry was already being equipped with a 3.7 cm anti-tank gun and there was the obvious desire for standardisation and economy. These requirements eventually resulted in the PzKpfw III and IV which are described on later pages. In the meantime 1932 saw the plans drawn up for the new Panzer Divisions, and the vehicle specifications were formulated. It was realised, however, that some time would necessarily be needed to develop the combat tanks. A training tank was required in the meantime which was inexpensive, could be rapidly developed, and could be built quickly.

A Carden-Loyd Mk IV tankette chassis was purchased from Vickers in England for study by the Army Weapons Office (Heereswaffenamt) in 1932 to assess its suitability as a carrier for a 2 cm anti-aircraft gun, but it was quickly apparent that this type of vehicle would make an excellent light training tank, albeit with only a machine gun armament. A requirement was

TOP TO BOTTOM: Krupp LKA 1. 1A La S Krupp (PzKpfw I Ausf A) in 1935. PzKpfw I Ausf A with infantry, Norway 1940.

issued by Heereswaffenamt for a 5 ton vehicle with twin machine guns in a fully traversing turret, and Rheinmetall, MAN, Krupp, Henschel, and Daimler-Benz were all invited to submit designs. Krupp's chassis design, which the firm designated LKA 1, was selected for development. This was influenced by the Carden-Loyd chassis. The responsibility for the superstructure and turret was given to Daimler-Benz.

In December 1933 an order for three running prototypes was placed, the Henschel firm of Kassel carrying out the construction. All this work was spread among several firms mainly for 'educational' reasons. The first vehicle was completed on February 3, 1934 and after successful trials, a contract was given to Krupp to produce 150 vehicles, starting in July 1934. The prototype LKA 1 had four road wheels supported on coil springs, and three return rollers. Production vehicles differed in having an external girder connecting the second road wheel and the idler, with quarter elliptic springs providing the suspension. Production vehicles were given the 'camouflage' designation 'Landwirtschaftlicher Schlepper' (abbreviated to La S), or 'agricultural tractor' to conceal the vehicle's true function. The first vehicles were fully designated '1A La S Krupp' when in service. It was soon found that the vehicle was susceptible to improvement, and a more powerful engine and longer hull with lengthened suspension (one extra road wheel each side) was provided. Subsequent production was of this improved type, initially designated '1B La S May'.

The La S was a full-track two-man vehicle with a fully traversing turrret offset to the right and carrying twin MG 13 machine guns – the turret was, in fact, similar to the auxiliary turrets later used on the big NbFz vehicles. The air-cooled Krupp M305 four-cylinder horizontally opposed engine was mounted at the rear and was of 3.5 litre capacity. The drive was taken via a dryplate clutch to a ZF Aphon FG 35 5-speed gearbox. Steering was of the clutch and brake type. Some vehicles were fitted for test purposes with a Krupp M601 air-cooled diesel engine but this failed to deliver sufficient power and was not pursued. In the improved 1B La S the Krupp engine was replaced by a water-cooled Maybach NL 38 TR engine of 100 hp at 3,000 rpm. An improved transmission was also fitted in this model. The driver sat alongside the turret to the left of the vehicle. The track was of the skeletal type, with dry-pins and twin guide horns. The armour thickness was 15 mm maximum.

The first La S vehicles were quickly in service, and these new vehicles made a big press impact when they took part in a military display at Doberitz in July 1935. A Panzer Training Division based on Munster-Lager was organised for that summer and the first three Panzer Divisions were officially formed in October that year. La S vehicles first saw real action in the Spanish Civil War where some were supplied to the Nationalists in 1936-37. It was never the intention that these should be anything more than training vehicles, but there were delays in producing the PzKpfw III and IV, and no less than 1,445 of these small tanks – redesignated Panzerkampfwagen I from February 1938 – were still in service when Poland was invaded on September 1, 1939, and over 800 were still in service in July 1941. They were thus used quite extensively in action in the early years of World War 2. While

Left and right views of PzKpfw I Ausf B. Vehicles of Panzer-Lehr Abteilung, 1938 (upper photo).

quite inadequate in firepower and armour protection, against the poor and fairly disorganised opposition of the time they gave quite good service. Five manufacturers were engaged in PzKpfw I production: MAN, Henschel, Krupp, Wegmann,and Daimler-Benz. Separate chassis number bands were allocated to each manufacturer, but, of course, the numbers were not necessarily allocated in full. Remaining vehicles in 1940-41 were converted either to self-propelled gun carriages or were used for training. The development sequence and further characteristics are described below.

LKA 1: Krupp prototypes, three only for trials, 1934.

1A La S Krupp/PzKpfw I(MG) Sd Kfz 101) Ausf A: Standard 1934 production vehicle to original design with Krupp air-cooled engine and short hull and suspension. Latter designation from February 1938. Chassis numbers: 8000-10477.

1B La S May/PzKpfw I(MG)(Sd Kfz 101) Ausf B: Introduced from late 1935, the balance of production was of this type. Lengthened hull and suspension were the distinguishing features. On this model the idler wheel was carried clear of the ground. The nose of the vehicle was slightly revised to take the modified transmision units, and the machine gun mantlet was of the internal type. Chassis numbers: 10478-16500.

PzKpfw I (Sd Kfz 101) Ausf A/B Flamm: A few vehicles of both the model A and model B type were converted, possibly in the field, to the flamethrower (Flammpanzer) role. In this conversion a flame gun replaced the right hand machine gun and fuel containers were carried in the turret. This type of vehicle was used in the Western Desert in 1941 and was probably also in action on the Russian Front.

PzKpfw I(A) Munitions-Schlepper (Sd Kfz 111): This was a 1939 conversion from the original tank by the removal of the turret. The result was a low close-topped vehicle (1.4 m high) which could carry ammunition either in the vehicle or in a light towed trailer. The driving position and controls were unaltered. This conversion work was carried out by Daimler-Benz, and totalled 51 vehicles.

PzKpfw I(A) Schlepper: This was a variation of the Munitions-Schlepper, issued in 1940 to some panzer units as a recovery vehicle to supplement or substitute for the 1 ton half-track tractor used mainly in this role. The early vehicles issued were on the Ausf A chassis, but as maximum power was required for a towing job, most subsequent conversions of this type were on the Ausf B chassis.

Fahrschulewagen I: This conversion was carried out on both Ausf A and Ausf B machines to suit them for driver training. The entire superstructure and turret was removed. Some vehicles of this type had a low rail round the open fighting compartment as a safety aid for men riding in the vehicle over cross-country terrain. NSKK (National Socialist Motor Corps) units were issued with vehicles of this type for training purposes.

Kleiner Panzer Befehlswagen (Sd Kfz 265)/K1 PzBfWg (Sd Kfz 265): To enable an armoured force commander to keep up with his troops, some Pz Kpfw I of both the A and B types were converted to command vehicles. There were three different models, 1 K1 B, 2, K1 B, and 3 K1 B, and most of the vehicles which actually saw service (a total of about 200) were on the B

PzKpfw I Ausf A Flamm fitted with Model 40 flame projector in turret (courtesy MAFVA).

CENTRE: PzKpfw I(A) Schlepper (at left), France 1940. ABOVE: Fahrsculewagen I of NSKK.

chassis. The earliest version, of which only a few were built, had a rectangular traversing turret, but this was dropped in favour of a rigid built-up superstructure in place of the original superstructure and turret. Deutsche Edelstahlwerke carried out the conversions. There was a crew of three, driver, radio operator, and commander. A map table and two radio sets, Fu 6 and Fu 2, were fitted, and there was an extra dynamo to keep the radio batteries charged. The armour plate on the superstructure front was 17 mm thick and an extra 10 mm was added on the nose. A ball-mounted 7.92 mm machine gun was carried in the superstructure front for self-defence. Some 96 vehicles of this type were used in the French campaign of 1940; they also served in the Western Desert and on the Russian Front in 1941.

Ladungsleger I: This was a project to convert PzKpfw Is for use as explosive charge layers (ladungsleger). In May 1940 a contract for the development of this vehicle was placed, calling for an extended arm mounted on the superstructure roof which could be used to place a 75 kg charge against a target, allowing the tank to withdraw before detonation. Several other experiments, including possible conversion to radio-control, were carried out with PzKpfw I demolition vehicles.

By the end of 1941 the PzKpfw I had almost disappeared from front line service, but the chassis served on for some time. Two self-propelled carriages were based on the chassis in 1940, the PzJg I (4.7 cm) and the Sf SiG 33 and these were produced in some numbers.

VK 601/PzKpfw I Ausf C: In September 1939 it was decided to further develop the PzKpfw I design both to undertake combat reconnaissance and to equip the airborne divisions which were then just being formed. This project was given the experimental designation VK 601, and Kraus-Maffei and Daimler-Benz were contracted to do the development work, basing the vehicle on the existing PzKpfw I Ausf B. The major changes were an increase in armour to 30 mm maximum, the replacement of the turret by a turret from the PzKpfw II tank, with 2 cm gun and co-axial 7.92 mm MG, and an uprated 150 hp Maybach HL 45 engine which gave an increased road speed up to 40 mph. This was facilitated by new interleaved suspension and wider track which gave improved flotation. The provisional designation PzKpfw I Ausf C was given to the vehicle and a pre-production order for 40 vehicles was placed; however, by the time the prototype had been delivered requirements were changed and a leisurely contract time was stipulated, asking for delivery by July 1942. As far as is known, however, this was later cancelled and the number completed is not known.

VK 1801/PzKpfw I nA/PzKpfw I Ausf D: In late December 1939 a further revised requirement was put out by the Weapons Dept asking for a prototype with the 'thickest possible armour' based on the PzKpfw I design. This type, designated VK 1801, was to supplement the VK 601 and the contract was given to the same firms. Armour of 80 mm thickness was incorporated which gave the hull and superstructure a much enlarged appearance. Side access hatches for the crew were fitted to the superstructure sides. The suspension and track was a strengthened verson of that used in the VK 601. The turret was of the same type as fitted to the standard PzKpfw I but with thick armour plating, and the twin MG 34s were in armoured

TOP TO BOTTOM: The rare early model Kl PzBfWg I with traversing turret, 1939. Standard Kl PzBfWg I on Ausf B chassis, France 1940. Kl PzBfWg I, late vehicle, Libya 1942 (IWM-BM7602).

sleeves. Engine and power train were the same as the VK 601 but the greatly increased weight (18-19 tons) reduced the speed to 25 kph (15 mph). This vehicle was probably inspired by the British Infantry Tank Mk I, then in service with the British troops in France, and a similar infantry support role was envisaged for it. The model was tentatively designated Panzerkampfwagen I neuer Art (nA = new model) and was later unofficially known as the PzKpfw I Ausf D. The prototype was ready in June 1940 and there was a pre-production order for 30 vehicles, a number of which were also delivered. An order for 100 production vehicles was cancelled. The PzKpfw I nA/Ausf D was used for training alongside various other VK models during the course of the war.

Bruckenleger I: This was a ramp-carrying conversion of the Ausf B, a number of which were in service for the campaign in France in 1940. A hinged end ramp was attached to the main ramp. One, two, or three vehicles could be used in combination depending on the gap to be spanned. The principle was subsequently used by the British under the general name ARK (armoured ramp carrier). Also used in small numbers was the Pionier-panzerwagen I, a standard vehicle with turret removed, for engineers' use.

Bruckenleger I. Two vehicles nose-to-nose forming an ARK-type bridge in France, June 1940, with PzKpfw II demonstrating a crossing. Note hinged access ramps, extended on left vehicle and stowed on right vehicle.

RIGHT: VK 601/ PzKpfw I Ausf C when new. BELOW: Front detail—note 'S' mine dischargers (anti-personnel) on mudguard (W. J. Spielberger). BOTTOM: VK 601 in 1944 with armament and outer road wheels removed.

TOP: VK 1801/PzKpfw I Ausf D. Note side hatch open (W. J. Spielberger). CENTRE: VK 1801 later in war with tac signs of an operational unit. ABOVE: Ladungsleger I showing charge-placing arm.

3: Panzerkampfwagen II

When it became obvious that development time for the main tanks required for the Panzer Divisions might be longer than at first anticipated, General Lutz decided that a slightly larger and more powerful training vehicle would be desirable. Thus the Heereswaffenamt in July 1934 asked three manufacturers to submit designs to meet a requirement for a 10 ton vehicle with 2 cm KwK 30 and a co-axial machine gun in a fully traversing turret. Krupp, MAN, and Henschel submitted designs, and Krupp quickly produced a prototype, which they designated LKA II, by converting the orignal LKA 1 to take a larger turret with the armament specified. However, the MAN design, which had a longer lower superstructure, was preferred, and prototypes were tested. As with other early designs, a 'camouflage' designation was given, La S 100 (Landwirtschaftlicher Schlepper 100 = agricultural tractor Type 100) until 1938 when the need for concealment was past and the designation Panzerkampfwagen II was applied. Between 1935-37 a number of pre-production series were built and these were extensively operated, the experience being used to introduce successive modifications and improvements. A major change took place in the suspension system and there were changes in engine layout and armour thickness.

Early models weighed 7.2 tons and later models 9.5 tons. The layout was conventional, with driver in the hull front and commander and gunner in the turret. The first model had a Maybach six cylinder HL 57 engine giving 130 hp at 2,100 rpm. There was a ZF six speed (five forward, one reverse) gear-box connected by a plate clutch to the cross-shaft which drove the front sprockets. There was clutch and brake steering for each track. Subsequent models had the uprated Maybach HL 62 engine, raising power from 130 to 140 hp. The two-man turret carried the 2 cm KwK 30 on the left and a co-axial MG 34 on the right. There was a hatch for the commander, vision ports, periscopes and a stowage box on the turret rear in later models.

Like the PzKpfw I, the PzKpfw II was not originally envisaged for combat use. However, it proved very reliable and effective. The PzKpfw II therefore remained in increasing production and was the most numerous German-built tank in service with the Panzer Divisions when the campaigns in Poland and France were undertaken in 1939 and 1940. Some 955 PzKpfw IIs were on strength when France was invaded in May 1940 and production continued until 1942 when over 800 were still on first-line strength. Latterly the PzKpfw II was used in the reconnaissance role, and saw service in the Western Desert and in Russia well into 1942. With no HE capability, and a very small calibre gun, the PzKpfw II was really outmoded in 1940. While it passed muster against fairly light opposition, as on the Western Front in 1940, it was quite severely decimated on the Russian Front in 1941-42.

Individual models are detailed and described separately below.

1/La S 100/PzKpfw II(2 cm)(Sd Kfz 121)/PzKpfw II Ausf al: First pre-production type with Maybach HL 57 engine, built in 1935. The road wheels were sprung bogies on leaf springs attached in pairs to a girder-type beam.

The nose plate was a rounded plate, and the gun had an internal mantlet. Chassis numbers: 20001 - 20025.

1/La S 100/ PzKpfw II Ausf a2(Sd Kfz 121): As Ausf al but with detail changes to the cooling system and engine compartment. Introduced in 1935, chassis numbers 20011 - 20025.

1/La S 100/PzKpfw II Ausf a3: A further pre-production series of vehicles built in 1936 and with improved tracks and strengthened suspension, plus further changes to the cooling system. Chassis numbers: 20026 - 21000.

2/La S 100/PzKpfw II Ausf b: The next batch of vehicles, built in 1936, were given the uprated HL 62 engine, improved tracks, and new drive gears and sprockets. The weight was increased to 7.9 tons, and the armour thickness on all these early series models was 14.5 mm. Chassis numbers: 21000 - 21100. Ausf b could be distinguished from Ausf a models by replacement of the rounded nose by extended nose plates over the new reduction gear, and an extended rear hull with vertical air outlet of rectangular grille pattern.

3/La S 100/PzKpfw II Ausf c(Sd Kfz 121): This model appeared in 1937 and saw fundamental changes to the design. The main alteration was the suspension, the girder and small bogies being replaced with large disc road wheels suspended on quarter-elliptic springs; this was used on all subsequent models. There were four return rollers. Detail changes were made to the turret and the front superstructure was widened slightly to give more room in the driving compartment. Weight rose to 8.9 tonnes. Chassis numbers 21101 - 23000. By this time several other plants had been brought into the Panzerkampfwagen II programme, and Famo, MIAG, Wegmann, and Henschel were engaged in production as well as MAN.

4/La S 100/PzKpfw II Ausf A(Sd Kfz 121): Mass-production started in a big way with the Ausf A. A new driver's visor was introduced on this model. Chassis numbers: 23001 - 24000.

5/La S 100/PzKpfw II Ausf B: There was little difference between these 1937 - 38 models. This model, and the Ausf C, were the main types in service in the early campaigns of World War 2. Chassis numbers: 24001 - 26001

6/La S 138 PzKpfw II Ausf D (Sd Kfz 121): In the attempt to produce a much faster vehicle, Daimler-Benz undertook in 1938 to develop a 'Schnellkampfwagen' (fast tank) version of the PzKpfw II with torsion bar suspension and larger road wheels. This vehicle had a completely new hull with angled front plates and a flat vertical superstructure front. Hatches, decking, engine covers, and other details were all virtually scaled down from PzKpfw III (the successful ZW 38 Daimler-Benz design). The crew was raised from 3 men to 4 men. The gearbox was a Maybach Variorex VG 102/28H with pre-selector, which gave 7 forward and 3 reverse speeds. The turret fitted was exactly the same as in the PzKpfw II Ausf A,B, and C.

7/La S 100/PzKpfw II Ausf C(Sd Kfz 121): The Ausf C was basically similar to the earlier La S 100 models but detail improvements included better engine cooling, vision ports, and a 50 mm armoured glass block for the driver in place of the 12 mm block formerly used. Production took place in 1938 - 39. Chassis numbers: 26001 - 27001.

8/La S 138/PzKpfw II Ausf E (Sd Kfz 121): The Ausf E remained similar

TOP: Krupp LKA II prototype. ABOVE: PzKpfw II Ausf a1. LEFT: PzKpfw II Ausf a2.

to the Ausf D, except that the track was changed to the rubber-padded needle roller bearing type normally associated with the German semi-track vehicle. Drive sprocket and wheels were modified to suit the track. The desired improvement in speed and handling was not realised in these models and the production was converted to flamethrower vehicles. Chassis numbers: 27801 - 28000.

Refitted vehicles: In the winter of 1939 - 40, following the campaign in Poland, a planned refit of all existing major production types was undertaken (ie, PzKpfw II Ausf c, A,B,C). The armour was increased to 35 mm by adding 20 mm plates over the existing 15 mm on all frontal surfaces. The original rounded front plate was thus concealed behind the additional sloped armour plates. From late 1940, further refits included the fitting of a new cupola as later appeared on the PzKpfw II Ausf F. These retrospective refits make the task of identifying individual models from photographs extremely difficult.

9/La S 100/PzKpfw II Ausf F(Sd Kfz 121): The final production model of the PzKpfw II was much improved to take into account actual battle experience and to enhance, as far as possible, the vehicle's reconnaissance qualities. Frontal armour of flat plates was increased to a basic 35 mm and side armour to 20 mm. The weight was increased accordingly to 9.5 tons and there was a slight fall-off in speed which was considered an acceptable sacrifice for the better protection. A new low cupola with seven vision blocks was fitted. Production started in March 1941 and continued to mid 1942. Chassis numbers: 28001 to 28500 (approx). Production of this final type of PzKpfw II continued well into 1942, despite the fact that this light tank was now completely outmoded. However, to expand the armour force to 36 divisions (as decided in July 1941) meant that there was a demand for all available tanks to make up the numbers and it was easier to keep established production lines going rather than switching plants to new types with consequent disruption of supply. Late production Ausf Fs had detail changes including a stowage box on a turret rear; in many cases they had a dummy driver's visor and dummy pistol ports added additional to the actual fittings, presumably to confuse the enemy.

PzKpfw II Ausf H and M(Sd Kfz 121): These were the final proposed variants of the PzKpfw II, but they were cancelled before entering production. Work started on the H model in September 1941 and it was to have a Maybach HL-P engine delivering 200 hp, 30 mm armour, wider tracks, and a larger three-man turret with 2 cm KwK 38 gun and MG 34. Top speed would be 65 kph(40 mph). Production was planned to start in July 1942 but plans for this were abandoned, and production of the previous PzKpfw II models was halted shortly afterwards.

PzKpfw II(Flamm)(Sd Kfz 122): Plans to convert some tanks to flamethrowers were drawn up in 1939 by Heereswaffenamt. A test series of 3 vehicles were built in 1939 and a batch of 85 of the unsatisfactory PzKpfw II Ausf D/E were converted to carry flamethrowing equipment starting in January 1940. All were used on the Russian Front. A second order for 150 similar conversions was cancelled in late 1941. The flame projectors were in small turret-like mountings, one each side at the front and traversing through 180 degrees. Some 80 flame shots of 2-3 seconds each were possible, and the

TOP TO BOTTOM: PzKpfw II Ausf C or A as built. PzKpfw II Ausf B after factory refit; note added front armour—model indentified by chassis number on front plate. PzKpfw II Ausf B or C after refit, France 1940 (IWM-MH4135).

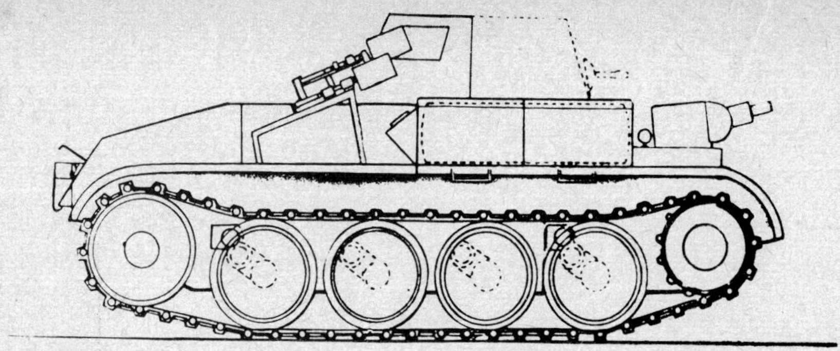

PzKpfw II (Flamm) (US Official).

range was 35 metres. The modified vehicle weighed 11 tons, had a two man crew and carried a single MG 34 in the turret. These vehicles were not very successful due to the limited range of the flame projectors, and those that survived were converted to self-propelled carriages.

Self-propelled carriages: The redundant PzKpfw II Ausf D/E and (Flamm) chassis were converted to Marder II with Russian 7.62 cm guns, and the PzKpfw II production chassis was kept going until 1944 to provide the basis for several types of motor carriage including the Wespe and Marder II. PzKpfw II tanks were also stripped and the chassis used for motor carriages.

Bruckenpanzer II: Magirus proposed a bridgelayer tank based on the PzKpfw II and a prototype using an Ausf b chassis existed. In France in 1940 some PzKpfw II carried a small bridge over the turret and engine compartment.

PzKpfw II Schwimmpanzer: For the projected invasion of England in 1940 (Operation Sea Lion) a special tank unit was formed (Battalion A) in Putlos and undertook training for amphibious assault by tanks from landing barges. Flotation chambers were to be attached to the return rollers allowing the vehicles to 'swim' ashore at 10 kph, with a propellor drive from the engine via a waterproof shaft. Flotation bags were placed in all available spaces inside the hull, the engine was sealed and waterproofed, and an inflatable rubber tube sealed the joint between turret and hull. The vehicle floated down to its track guards and could operate in up to winds force 3-4. Over 50 sets of flotation gear were ordered.

VK 901/PzKpfw II nA(PzKpfw II Ausf G): An early attempt to give a faster and better armoured version of the PzKpfw II started development in June 1938 when MAN and Daimler-Benz were asked by Heereswaffenamt to produce a model with 30 mm frontal armour within a weight limit of 10 tons and with a speed of 60 kph. The prototype, VK 901, was completed at the end of 1939. It had widened track and strenghtened overlapped suspension similar to the VK 601 (PzKpfw I Ausf C). The turret was of the standard PzKpfw II type but was fitted with a 2 cm KwK 38, an extended mantlet, and a stabiliser, in place of the 2 cm KwK 30. The engine was a Maybach HL 45 of 145 hp, which gave a speed of 50 kph, falling short of the required maximum. A 200

TOP: PzKpfw II Ausf D (with PzKpfw I Ausf A nearest). ABOVE: PzKpfw II (Flamm), rear view. BELOW: PzKpfw II (Flamm), front view.

hp engine was necessary but was not available at that time. Some 75 pre-production vehicles were ordered to be built, starting in October 1940. Only limited numbers were completed in the 1941-42 period.

VK 1601/PzKpfw II nA (PzKpfw II Ausf J): Another prototype ordered from MAN and Daimler-Benz in December 1939 was to have the 'thickest possible armour' and was a parellel development to the VK 1801(PzKpfw I Ausf D), sharing many of its characteristics. The vehicle was massively constructed, with 80 mm frontal armour and 50 mm at the sides. Wide tracks and heavy overlapped suspension were fitted, and there were side access/escape hatches for the three-man crew. Armament was as for the VK 901. Weight was 16-17 tons and a top speed of 31 kph was given by the Maybach HL 45 150 hp engine. There was a pre-production batch of 30 vehicles ordered in June 1940 and the first delivery was scheduled for December 1940. An order for 100 further vehicles was abandoned. The vehicles, completed during 1941 and 1942, were later used for training and experiments, and some were converted to recovery vehicles.

Bergepanzer II(PzKpfw I nA)(VK 1601): At least one, possibly more, of the VK 1601 vehicles was converted to a recovery tank with a girder-built gantry and winch replacing the turret.

VK 903/PzKpfw II nA: A proposal was made in June 1942 to develop the existing VK 901 into a specialised armoured reconnaissance vehicle for armoured artillery and panzer regiments. The turret from the VK 1303 (Lynx prototype) was to be used, along with the driver train of the PzKpfw 38(t)nA. The vehicle was also fitted with rangefinder, sound locator and other specialised command equipment. It did not proceed beyond the prototype stage. It is possible that the PzKpfw II Ausf H or M designation was earmarked for this design.

VK 1301/PzKpfw II nA: Drawing on the result of trials with this VK 901 and VK 1601, Heereswaffenamt asked MAN and Daimler to produce a definitive armoured reconnaissance vehicle combining the best features of the two. A mild steel prototype was built in April 1942. In essence the VK 1301 was a slightly enlarged version of VK 901, with overlapping wheels and revised superstructure. Two further prototypes were built, VK 1302 and VK 1303, with detail changes aimed at reducing the weight. The VK 1303 was selected for production, becoming the Lynx(Luchs).

PzKpfw II Ausf L Luchs (Sd Kfz 123)/Panzerspahwagen II Luchs: This was the full production vehicle derived from the VK 1303 as a specialised reconnaissance type. A Maybach HL 66P engine of 180 hp gave a top speed of 60 kph and the weight was 11.8 tons. The overlapping disc wheels had torsion bar suspension and there was plentiful stowage for stores. Classed originally as a tank, it was later redesignated as a Panzerspahwagen (armoured recce vehicle). There was a four man crew. The armament in the first 100 production vehicles was to be 2 cm KwK 38 gun with co-axial MG 34. The very limited production was terminated in January 1944, but the type was in service until the war's end. Armour was 30 mm in front and 20 mm at the sides. Later vehicles were to be fitted with a 5 cm gun.

VK 1602/Gefechtsaufklarer Leopard: This project was started by Heereswaffenamt in 1941 to produce a tracked vehicle suitably armed for bat-

TOP TO BOTTOM: PzKpfw II Ausf F; note dummy driver's visor (IWM-STT4625) PzKpfw II Schwimmpanzer. Bruckenleger PzKpfw II Ausf B, Magirus prototype (Werner Oswald).

tle reconnaissance — ie, well enough armed and armoured to engage enemy vehicles if necessary. Armour of 50 - 80 mm, on the turret and 20 - 60 mm on the hull, and a 5 cm KwK 39/1 gun with co-axial MG 42 were the key features. A 550 hp engine was to give a top speed of 60 kph. MIAG and Daimler-Benz undertook development, working on the design (which resembled VK 1301) through 1942. After the prototype was built the project was abandoned in 1943, but the turret design was adapted for the Puma armoured car.

Miscellaneous conversions: Munitionspanzer II, Pioneerpanzerwagen II PzKpfw Fahrschulewagen II — all turretless vehicles for ammunition supply, engineer stores, and driver training respectively. PzBef Wg II — command vehicle with extra radio.

BELOW: Bruckenleger II; service model, France 1940. BOTTOM: VK 901 (PzKpfw II Ausf G).

TOP TO BOTTOM: VK 1601 (PzKpfw II Ausf J). VK 1301. Pioneer - panzerwagen II—turretless conversion for engineer use.

TOP: VK 1601 (PzKpfw II Ausf J) in service at an anti-tank school, 1944. ABOVE: Bergepanzer II (converted VK 1601), four views (Courtesy MAFVA). OPPOSITE, BOTTOM: Early production PzKpfw II Ausf L, Luchs.

LEFT: VK 1303 (Luchs prototype) on test. BELOW: Late production PzKpfw II Ausf L, Luchs (ECA).

VK 1602/Gefechtsaufklarer Leopard prototype.

PzKpfw II Ausf C in typical uparmoured and refitted form, North Africa 1941 (IWM-E2685).

4: Ex-Czech Tanks

NOTWITHSTANDING Germany's ambitious plans for the development of tanks, the production programme, especially for the planned main fighting tanks—PzKpfw III and IV—lagged well behind schedule and in the immediate pre-war years and right up to the campaigns in Poland and France, there was far too much reliance put on the light training tanks, PzKpfw I and II. The occupation of Czechoslovakia by German forces in March 1939 had the incidental but very important effect of placing all the equipment of the Czech armed forces, plus the vigorous and enterprising Czech arms industry, in German hands. There were two major Czech tank types which were technically superior to both the PzKpfw I and Pzkpfw II, and were also better armed. The measure of the importance of this windfall to the Germans can be shown by the fact that about 25% of all tanks in German service in the 1940-41 period were Czech tanks, and that three Panzer Divisions in this period (a notable one was 7.Pz.-Div. commanded by Rommel in May-June 1940) had these Czech tanks forming most of the vehicle strength. Arguably the German Panzer Division successes of the 1939-40 period owed much to the availability of the Czech tanks. The armoured divisions would certainly have been less formidable and more severely curtailed without them. The Czech factories were harnessed to the German war machines and production of tanks (and other armaments) was carried on for the Wehrmacht. German-inspired production changes were instituted and new designs were developed. The overall importance of these Czech resources to the Germans must, therefore, be emphasised. The tanks concerned, and the models developed, are summarised here.

PzKpfw 35(t)

Developed in 1934-35, and in production in 1935-38, this tank was a Skoda design (S-II a, later known as the T-11) adopted by the Czech Amy as the Lt Vz 35. It weighed 10.5 tons and had a maximum armour thickness of 25 mm. Armament was a 3.7 cm KwK 34(t) gun with excellent control characteristics and a high velocity performance. There was a co-axial 7.92 mm machine gun, and another machine gun in the hull. For its period this was a mechanically refined vehicle, with a top speed of 40 kph, a 12-speed gearbox with an air-assisted mechanical steering unit which made driving and control particularly easy and allowed long 'marches'. The engine and sprockets were at the rear, simplifying the layout, while the vehicle was of riveted and bolted construction. Bogie suspension was excellent and track life was long. Taken into German service in 1939, the vehicle was redesignated and deployed operationally as the PzKpfw 35(t). (t) was the abbreviation for Czechoslovakia in German. Over 200 were taken into service, most being issued to 6.Pz.-Div. They were used in both the Polish and French, and later even the Russian campaigns; in the last theatre cold weather had a detrimental effect on the air-assisted steering units, necessitating the use of heaters. Some had a fully mechanical steering unit installed. From late 1942 they went out of first-line service and were used for training or were converted to other roles. The Slovak and Romanian Armies, allies of the Germans, also used this tank, which they acquired from Skoda before 1939.

PzBefWg 35(t): This was the standard tank converted to act as a command vehicle with the addition of radio equipment for the purpose. It was identified by the large frame antennae over the engine compartment. A small number were issued to units equipped with the PzKpfw 35(t), eg, 6.Pz.-Div.

Zugkraftwagen 35(t): This was the PzKpfw 35(t) with the turret removed for use as a tractor or armoured recovery vehicle (tower) for light vehicles. It could tow a 12 ton load and was a conversion from redundant tanks.

Morserzugmittel 35(t): Similar to the tractor, but with more of the superstructure roof cut away, this towed a heavy wheeled mortar and gave accommodation to the crew and ammunition.

PzKpfw 38(t)

Much the most important Czech tank, the TNHP-S was designed by the CKD firm (Ceskomoravska Kolben Danek) and was to enter production for the Czech Army in 1938. Prior to adoption by the Czech Army it had been developed through several export models which had been supplied to other countries including Switzerland and Iran. The great virtue of the design was its compactness and immense reliability—on test for the Czech Army it ran 3,000 miles without breakdown. Maintenance was simple too. The vehicle weighed eight tons, was of riveted construction, mounted a 3.7 cm KwK 38(t) gun a co-axial 7.92 mm machine gun, another machine gun in the superstructure front, and had up to 25 mm armour—more in later models. Power was from a Praga 7.7 litre six cylinder petrol engine, and top speed was 42 kph. Drive was to the front sprockets via a Praga-Wilson epicyclic preselector gearbox with five speeds. Four large road wheels on longitudinal leaf springs and mounted in pairs gave an excellent ride. Turret could be traversed by gearing or manually. There was a crew of four, commander, gunner, driver, loader/radio operator.

PzKpfw 38(t) (3.7 cm): (Sd Kfz 140): This was the designation given to the TNHP-S in German hands. Some 150 were initially ordered from the stock of parts already in the Prague works of CKD. They played an important part in the Polish campaign and there were 228 equipping 7.Pz-Div. and 8.Pz.-Div. for the French campaign of 1940.

The vehicle was much liked and production was continued at about 40 tanks a month initially. The CKD now under German control, was renamed Bohmisch-Mahrische Maschinenfabrik. Some 1,410 tanks were built up to 1942, after which the chassis was used only as a self-propelled carriage. There were several different production models.

Ausf A: Original production order of 1939. Certain changes were incorporated to meet German specifications, such as increase of crew from 3 to 4 men. Armour maximum was 25 mm. Chassis numbers: 0001-0150.

Ausf B: Further production order placed by the Germans. Chassis numbers: 0151-0260.

ABOVE: PzKpfw 35(t) and vehicles in service, France 1940 (BA-269-228-14).
BELOW: PzKpfw 35(t) as received from Czech Army.

1

ABOVE: Morserzugmittel 35(t). Zugkraftwagen 35(t) was similar but with full superstructure roof intact. BELOW: PzKpfw 38(t) Ausf A.

Ausf C: Further production order for German service. Chassis numbers: 0261-0370. There was little external difference in these first three production batches. The first Ausf C vehicles were delivered in time for the campaign in France.
Ausf D: Photographic evidence shows a model with a straight driver's front plate of 25 mm thickness and this is assumed to be the Ausf D. Some vehicles are shown to have the exhaust pipe position raised, presumably to allow better clearance for rear smoke emitters. Chassis numbers: 0371-0475.
Ausf E: Introduced in the fall of 1940, like the PzKpfw III Ausf H and PzKpfw IV Ausf E, this was an uparmoured model. Riveted construction of this tank makes recogniton of individual models very difficult. Frontal surfaces were increased to 50 mm thick armour by rivetting two 25mm thicknessess together but externally there was little apparent difference. The sides were increased to 30 mm thick armour plating (15 + 15 mm) in a similar fashion. In some vehicles a German 3.7 cm gun was fitted. The weight went up to 11 tons. New storage arrangements included smaller/storages boxes and a tubular box for gun cleaning gear. Chassis numbers: 0476-0750.
Ausf F: Produced in 1941, this was essentially similar to the Ausf E but lacked provision for rear smoke emitters. Chassis numbers: 0751-1000.
Ausf S: In the spring of 1940 the Germans confiscated the 90 tanks which were under construction for an order negotiated between CKD and Sweden before 1939. The components were held, and after an agreement was made with Sweden for license production of the PzKpfw 38(t) (Swedish designation: Str v M/41) in December 1940, the go-ahead was given for BMM to complete the original 90 export vehicles for German use. The model designation 'S' stood for Sweden. Chassis numbers: 1001-1090.
Ausf G: Built from late 1941, this was similar to models E and F but with the basic armour increased to single plates of 50 mm and 30 mm thickness rather than the two plates riveted together as previously used. There were other minor technical changes. Not all Ausf G were completed as tanks, as from April 1942 all PzKpfw 38(t) chassis production was switched to self-propelled gun production. Chassis numbers: 1101 onward.

PzBefWg 38(t): This Panzerbefehlswagen variant was produced as a commander's model to serve units equiped with the PzKpfw 38(t), eg, 5.Pz.-Div., which used the PzKpfw 38(t) in Russia. Extra radio equipment was carried and in some cases the front machine gun was removed to give more internal room. Others were distinguished by a large frame antennae mounted over the engine decking. Some vehicles of this type were used later as command and observation vehicles for self-propelled gun units.

Aufklarungspanzer 38(t) (Sd Kfz 140/1): This was a conversion of the original tank design carried out in 1944 to provide a fast tracked reconnaissance vehicle for the Russian Front where terrain conditions in the winter limited the use of wheeled armoured cars. The turret and superstructure was removed, and a new superstructure incorporated an open-topped turret from the Sd Kfz 222 armoured car. A 2 cm KwK 28 and a MG 34, formed the armament. Integral stowage boxes, accessible from the fighting compartment, were built on to the superstructure. Some 70 vehicles of this type were converted in 1944.

TOP: PzKpfw 38(t) Ausf A or B (W. J. Spielberger). ABOVE: PzKpfw 38(t) Ausf E or F. OPPOSITE, TOP: PzKpfw 38(t) Ausf E. Note raised rear exhaust pipe and turret stowage box.

Munitionsfahrzeug auf Fahrgestell PzKpfw 38(t): Standard PzKpfw 38(t) with turret removed, used for utility service as a gun tractor and munitions carrier from 1943 on.

PzKpfw 38(t) Fahrschulewagen: From 1943 many old PzKpfw 38(t) were retired from front line use and saw service with training schools.

PzKpfw 38(t) mit Holzvergaser: A small number of tanks had the turret removed and were fitted with wood gas generating equipment, in order to save fuel.

PzKpfw 38(t) nA: Designed in 1941 (nA = neuer Art = new model), this was the BMM answer to the OKH and Heereswaffenamt requirements for a reconnaissance tank. Other designs for the same requirement produced by German industry were the VK 901, VK 1301 and VK 1601, etc. A new welded hull and turret were designed for the new model but the armament remained the same. A 250 hp V-8 engine was used. A limited number of trials vehicles were built, but like the German designs, the PzKpfw 38(t) nA was cancelled when the requirement for this type of vehicle was dropped in 1942.

PzKpfw T-15: Other than their PzKpfw 35(t) which was the previous Czech Army tank, Skoda had not been involved in the production of tanks for the Wehrmacht. However, its tank design department was still very active and several prototype and development vehicles were constructed between 1939 and 1941. The Hungarian medium battle tanks and assault guns were all based on the Skoda T-22 design. Skoda offered the T-15 design to Heereswaffenamt in 1941 to meet the same requirement as the PzKpfw 38(t) nA. Intended as a fast reconnaissance vehicle, it was to have a wide, long chassis, not unlike that of the PzKpfw 38(t). It featured a 220 hp air-cooled diesel engine, 50 mm frontal armour, and 30 mm at the sides. There was a well-sloped glacis plate and there was provision for mounting a gun up to 5 cm calibre. Weight was 10.5 tons and top speed 38 mph. A wood mock-up was built, and an actual prototype vehicle may have been completed in 1942.

PzKpfw T-25: Heereswaffenamt had a requirement for a tank in the 25 ton class, and Skoda offered their 7.5 cm KwK T-18 design as a basis for this. It

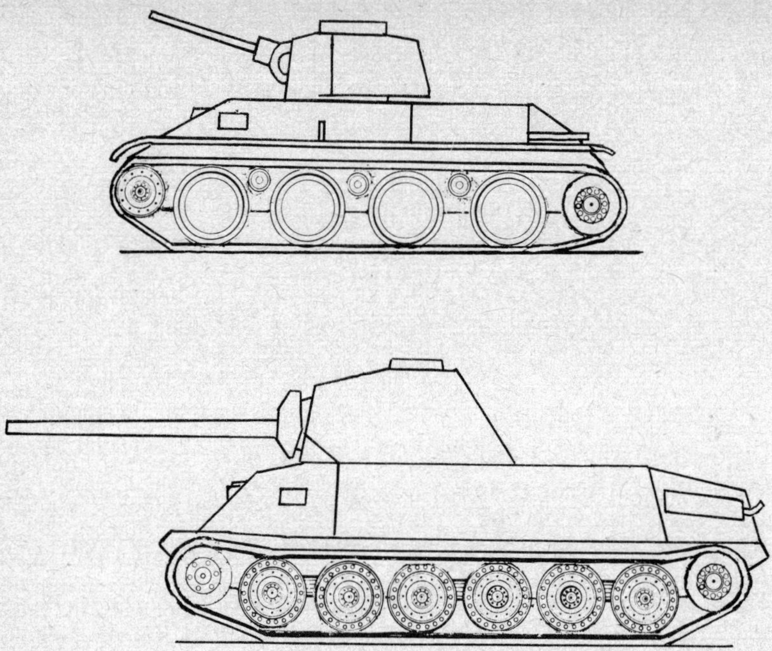

Impressions from orginal project drawings of PzKpfw T-15 (top) and T-25.

was to have automatic loading and a performance similar to the L/70 7.5 cm gun in the Panther. The T-25 superstructure was to be fully sloped and angled, similar in style to Soviet T-34 which inspired the Heereswaffenamt request. The vehicle was to be of all-welded construction with a 450 hp air-cooled Skoda engine. The automatic gun would have a drum type magazine. No prototype was built, the Panther tank being preferred, but the design was completed. The chassis design was also to be used in the Heuschrecke Waffentrager programme with a 10.5 cm gun, but only a mock-up of this was completed.

Projects: In an attempt to create an up-gunned version of the PzKpfw 38(t), Krupp proposed in 1944 to fit the turret from the PzKpfw IV with 7.5 cm KwK 40 L/48. However, this was a questionable design and the project was abandoned. A further project in late 1944 envisaged using the 7.5 cm KwK 42 L/70 on the Panther tank in a simplified shielded mount with a limited traverse. It was never proceeded with.

Self-propelled mounts: The PzKpfw 38(t) chassis was used extensively as a basis for several major self-propelled types. From mid 1942 all output of this chassis was used exclusively for self-propelled equipment. Marder III, Bison, and later the Hetzer were types of major importance based on this chassis. A modified version of the chassis design featured prominently in the Waffentrager programme.

TOP: PzBefWg 38(t)—note frame aerial above engine. ABOVE: Munitions-fahrzeug auf Fahrgestell PzKpfw 38(t).

TOP: Aufklarungspanzer 38(t)—note widened superstructure and armoured car turret. ABOVE: PzKpfw 38(t) nA.

5: VK 2001 Prototypes

THE genesis of what were to become both the PzKpfw III and the PzKpfw IV designs, the principal tanks of the Panzer Divisions, stemmed from a policy meeting on January 11, 1934, by the various departmental heads of the Heereswaffenamt (Wa Pruf 6). This meeting was to determine the best balance of AFVs for the new 63 division army which was being planned. The influence of General Lutz, Inspector General of Motorized Troops, and his Chief of Staff, Guderian, was considerable. Their ideas for a 'light' tank with a 5 cm armour-piercing gun and a 'medium' or 'close support' tank with a low velocity 7.5 cm gun, were upheld (Lutz was also head of Wa Pruf 6 at this time). Also accepted was the idea of a Motorised Troops Command to take charge of development and training, and Lutz and Guderian became C-in-C and Chief of Staff respectively of this command when it was formally established in June 1934.

For the new tanks there was to an absolute weight limit of 24 tons which was to keep them within existing road bridge limitation in Germany. There was an early compromise on the armament of the 'light' tank. Though Lutz and Guderian postulated a 5 cm gun, the Weapons Department and the Artillery wanted to settle for a 3.7 cm gun, as a weapon of this calibre (Pak 35/36) was already in production for use in the field by infantry anti-tank companies. In the interests of economy the Motorized Troops agreed with this, with the important proviso that the design should include a turret and turret ring big enough to take the larger calibre gun, if necessary, later. As it happened this was a crucial decision which went a long way towards alleviating the choice of the smaller calibre gun a few years later when the 5 cm gun became a necessity.

The January 1934 meeting led to the decision to develop and arm new Grosstraktor prototypes, which became the NbFz/PzKpfw V and the NbFz/PzKpfw VI models (described previously), the 'close support' tanks which were originally referred to as Mittleren Traktors (medium tractors), and the 'light' tanks which would be a new generation of the Leichtetraktor (light tractor). The most fundamental difference between the new generation of 'close support' (or 'medium') tanks and the 'light' tanks was in the armament, and other requirements originally formulated, such as speed, suspension, crew numbers and disposition, vehicle layout, weight limits, and so on were similar.

These various requirements were expressed in the project designation VK 2000 which was put out to various leading arms producers and engineering firms. (VK 2000 = Vollkettenkraftfahrzeug 2000 = fully tracked experimental vehicle of 20 tons). At about this time (1934) the 'camouflage' designations, BW (Bataillonsfuhrerwagen = battalion commander's vehicle) was adopted for the 'close support' tank and ZW (Zugfuhrerwagen = platoon commander's vehicle) was adopted for the 'light' tank.

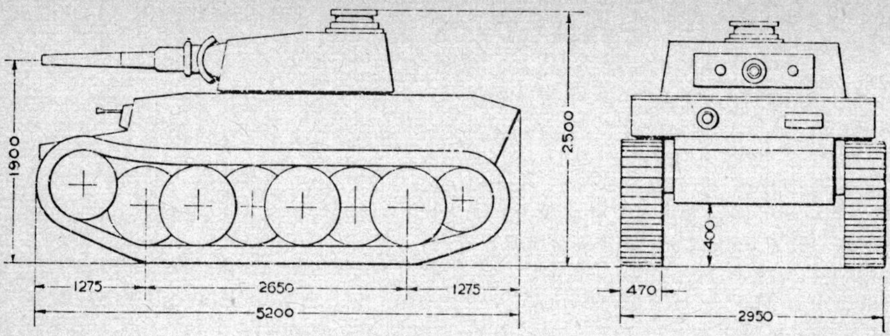

Dimensions in millimetres

VK 2001 (K)

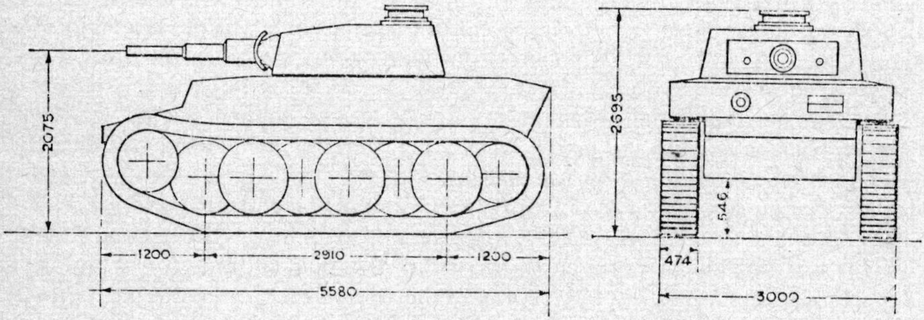

VK 2001 (MAN)

The submitted designs were as follows:

VK 2001 (Rh-B): This was the Rheinmetall-Borsig prototype, which was actually built in mild steel form and tested at the Kummersdorf proving ground, being completed at the end of 1934. Despite the official specification calling for interleaved suspension, the VK 2001 (Rh-B) had the standard Rheinmetall suspension as used in their version of the Grosstraktor and in their commercial crawler tractors. This consisted of four twin bogies each side suspended from sprung beams. There were three return rollers each side and mud chutes, plus front sprocket drive. The turret was without a gun.

VK 2001 (K): This existed only as a design project and did not get built. The drawings submitted to Wa Pruf 6, however, showed it to fit exactly to the letter of the VK 2000 requirement as regards turret, suspension, and layout. The original drawing shows a L/70 gun rather than the short L/24 gun which was eventually adopted for the PzKpfw IV. The turret from this design was, however, that adopted (with minor changes) for the actual production version of the PzKpfw IV, Krupp winning the production contract for turret and armament.

VK 2002 (MAN): This was MAN's design, submitted but not taken up. It was both longer and higher than Krupp's VK 2001. No order for a running prototype was given.

MKA: None of these VK 2000 series prototypes was evolved directly into

TOP: VK 2001 (Rh-B) (W. J. Spielberger). ABOVE: Krupp MKA prototype.

the BW tank. Such was the interchangeability between the BW and ZW requirements that a prototype submitted by Krupp the next year, 1935, for the ZW design was actually used as the main basis for the BW design, the entire development and production contract going to Krupp. The prototype was designated MKA by Krupp and had all the features which subsequently became familiar on the PzKpfw IV, notably hull and superstructure shape, suspension, and general layout. The MKA had a 3.7 cm gun to meet the ZW requirement; to evolve into the PzKpfw IV/BW design, the MKA design was adapted to take the turret already designed for the VK 2001 (K). Instead of an interleaved suspension there was a much simplified leaf spring bogie suspension with the road wheels in pairs. While the MKA had three bogies each side and three return rollers, the BW design which was evolved from the MKA was slightly longer, having four bogies each side and four return rollers.

6: Panzerkampfwagen III

THOUGH events dictated that the small training tanks, PzKpfw I and II, and the captured Czech tanks, were used extensively in the first two years of World War 2, it was the intention that the Panzer Divisions should be fully equipped with vehicles designed to accord with the ideas formulated by Lutz and Guderian. The three light companies of a tank battalion were to have a tank with a 5 cm armour piercing gun, a co-axial turret machine gun, and a front hull machine gun. This would thus become the principal combat tank of the Panzer Divisions. While discussions were taking place an early compromise was made in 1934 over the question of armament. While the Motorized Troops wanted the 5 cm gun, the Weapons Department and the Artillery wanted to settle for a 3.7 cm gun as a weapon of this calibre (Pak 35/36) was already in production for use in the field by infantry anti-tank companies. In the interests of standardisation the Motorized Troops agreed to the smaller calibre gun but with the proviso — of great importance later — that the turret and turret ring of the new design be big enough to accept a larger calibre gun.

During 1935, Heereswaffenamt put out outline specifications for the new vehicle. While an absolute weight limit of 24 tons (metric) was imposed due to bridge load restrictions in Germany, the actual requirement in this case was for a 15 ton vehicle. It was to have a top speed of 40 kph, main gun with co-axial machine gun, bow machine gun, a five man crew (driver, radio operator/hull gunner, commander, gunner, and loader). The commander was to have an all-round vision cupola and was to be centrally stationed in the turret. All vehicles were to be radio fitted, and throat microphones were to be used — a big advance in communications ideas for the time. The 'camouflage' name of Zugfuhrerwagen(ZW) — 'platoon commander's vehicle' — was used for the project to conceal its true purpose. MAN, Krupp, Rheinmetall-Borsig, Henschel, and Daimler-Benz were all asked to tender designs, and that of Daimler-Benz was selected for development and production from three actual prototypes which were produced by DB, MAN, and Henschel. Tests of the DB prototype were carried out at Ulm and Kummersdorf proving grounds from late 1936 through to the end of 1937. The specification given to DB by Heereswaffenamt was actually an amalgam of the best features of all three designs.

To finalise design features a series of pre-production vehicles was turned out in small batches for troop trials and the various models are detailed below.

Typ 1/2W/PzKpfw III Ausf A: Ten of this original model, were produced in 1937. The weight was 15.4 tons, armour from 5 - 14.5 mm, and armament was one 3.7 cm KwK L/35 with two MG 34 machine guns in a co-axial mount with an internal mantlet. There was another MG 34 ball-mounted in the front superstructure. The cupola was of the plain drum type with eight vision slits. There were single side doors on the turret, square pistol ports on the turret

TOP: Pilot model, PzKpfw III Ausf A. CENTRE: PzKpfw III Ausf A, Poland 1939. ABOVE: PzKpfw III Ausf B, Poland 1939 (BA-12/16/17).

rear, while the driver's visor in the front superstructure was a single flap hinged at the top. The engine was a Maybach HL 108 TR of 250 HP. Top speed was 32 kph. The suspension was comprised of five independently sprung bogie wheels on each side. Each wheel was connected to an arm which was in turn connected to the hull shaft of the bogie wheel. A coil spring was attached to the middle of the suspension arm. The driving sprocket was of disc type with eight perforation holes, and the idler had eight thick spokes. Some 120 rounds of 3.7 cm ammunition and 4,425 round of MG 34 ammunition could be carried. Chassis numbers: 60101 - 60110.

Typ 2/ZW/PzKpfw III Ausf B: Built in 1937, this had a new cupola, redesigned engine cooling outlets (ventilating upwards) and an alternative design of suspension featuring small wheels (8 each side) paired on bogies, each bogie being suspended by a semi-elliptic leaf spring, each with a separate shock absorber. This new suspension arrangement was intended to give a smoother ride and overcome problems experienced at speed with the Typ 1/ZW. Extra armour plate was fitted to the 2/ZW, notably round the tail. The cupola was the same as that developed for the early BW(PzKpfw IV) models. Fifteen vehicles were built, chassis numbers: 60201 - 60215. This 2/ZW chassis formed the basis of the first 7.5 cm StuK auf Fgst PzKpfw III (StuG III), five vehicles being built in 1937 - 38 for trials. Round hatches on the nose plate were a feature of some, if not all, 2/ZW vehicles.

Type 3a/ZW/PzKpfw III Ausf C: Also built in 1937-38, concurrent with Typ 2/ZW, this model had a further suspension variation to give eight small road wheels each side, the two inner pairs connected by semi-elliptic leaf springs, with the forward and rear pairs mounted on individual semi-elliptics. There were three return rollers. In all other respects it was the same as Typ 2/ZW. Production total was 15. Chassis numbers: 60301 - 60315.

Typ 3b/ZW/PzKpfw III Ausf D (Sd Kfz 141): In production in 1938, this was the first model to see quantity production, 30 being built alongside 30 command vehicles of the same type. The suspension was similar to that of Ausf C except that the semi-elliptic springs of the front and rear road wheels were angled upwards. While Ausf A-C had two fuel tanks, Ausf D and two extra ones added increasing tankage from 33 gallons to 66 gallons. Armour maximum at the front was increased to 30 mm and a new armoured cupola was fitted. Chassis numbers: 60221 – 60255 and 60316 – 60340

The command vehicles (Panzerbefehlswagen) produced concurrently were designated Typ 3c/ZW. When the Typ 3b/ZW entered service, some of the earlier Typ 2/ZW and 3a/ZW vehicles were re-worked roughly to Typ 3b/ZW style with consequent changes to the suspension, engine, armour and cupola, and these were then also known as PzKpfw III Ausf D. Official German records show that the front armour for the Typ 3b/ZW was increased to 30 mm, but photographic evidence shows that at least some of these vehicles retained the original 14.5 mm armour; it is possible, however, that the vehicles with the thinner armour were from the chassis group 60221 - 60225 (which were originally to be Typ 2/ZW vehicles) or were Typ 2/ZW and Typ 3a/ZW machines modified to Typ 3b/ZW standards without an increase in armour thickness. In the Polish campaign of September 1939 examples of all the early ZW models were employed, Typ 1/ZW (Ausf A) in its original form

LEFT: PzKpfw III Ausf B. BELOW: PzKpfw III Ausf D. BOTTOM: Rear detail of PzKpfw III Ausf E.

and Typ 2/ZW and 3a/ZW either in original or modified forms. In September 1939 and PzKpfw III designation was formally adopted for the ZW (at which time the Ausf A,B,C,D models were so designated) and the ordnance number Sd Kfz 141 was allocated.

ZW 38/Typ 4/ZW/PzKpfw III Ausf E: In 1938 Daimler-Benz had produced the design for a greatly improved model, under the project designation ZW 38. To overcome the suspension problems experienced in some degree or other with all the early ZW models, a new torsion bar suspension was developed. This work was done with the personal encouragement of Ing Kniepkamp, head of Wa Pruf 6 and in the same year DB gained experience in this type of suspension with their DB9 12 ton semi-track, and the PzKpfw II Ausf D and E. Subsequently the torsion bar suspension system was adopted as standard for the ZW(PzKpfw III) series. Other features of the ZW 38 included greatly increased rear and frontal armour of 30 mm, with consequent re-design of the driver's visor and hull machine gun mount. The driver was given an improved visor with an upper and lower sliding shutter, and the hull machine gun was set behind a square plate which was bolted to the superstructure front. The extra weight of the heavier armour had an effect on performance, and it was decided to introduce a more powerful version of the Maybach engine, the HL 120 TR V-12 of 11.9 litres. This developed 320 hp at 3000 rpm. A new pre-selector syncromesh gearbox, Variorex SRG 328-145, was fitted, having ten forward speeds, two of them overdrives, and four servo-operated reverse speeds. The suspension consisted of six road wheels each side, each independently sprung on a torsion bar which was carried across the hull. A radius arm connected the wheel to the torsion bar which was set ahead of the wheel. To allow the bars to fit, the wheels on each side had to be offset from each other by about 5 inches. There was a shock absorber on the first and last road wheel each side. The idler wheel on this model was changed to a ribbed type of welded construction, but the sprocket remained unchanged, and there were three return rollers. As Typ 4/ZW, construction totalled 96 vehicles and early vehicles produced saw service in the Polish campaign. Chassis numbers: 60401 - 60496.

Typ 5/ZW/ PzKpfw III Ausf F(Sd Kfz 141): This model went into production in October 1939 and was identical in most respects to the Ausf E with the addition of an anti-splash rail on the superstructure roof to protect the turret ring. Brake air intakes on the nose were the main external difference introduced on the Ausf F. These vehicles were used in the campaign in France in 1940. Many Ausf F were later upgunned to take the 5 cm KwK L/42 and increased armour in subsequent refits. Standard fitting on all early models was smoke emitters on the hull rear. These were carried in a rack on the right side and five generators were installed. They were held by spring-loaded catches and released by a rod linkage from the turret, working on the ratchet principle. Chassis numbers: 61001 - 65000.

It was realised early on in the PzKpfw III production programme that the original decision to standardise on the 3.7 cm gun was a mistake, and in 1938 Krupp were asked to develop a new mounting and a 5 cm gun suitable for the ZW vehicle. It was 1940 before this new mount was ready for production and the Ausf G was the production vehicle first fitted with this new weapon. The

TOP: Detail plan view of PzKpfw III Ausf E—note internal mantlet (Axel Duckert). ABOVE: PzKpfw III Ausf G with 5 cm gun. BELOW: PzKpfw III Ausf G followed by Ausf H (IWM-GER621).

gun was a 5 cm KwK L/42. The 5 cm gun had an external mantlet, was electrically fired, and had a co-axial MG 34 in the mount. A retrospective addition was the fitting of a stowage box on the back of the turret in later vehicles.

Typ 6/ZW/PzKpfw III Ausf G (Sd Kfz 141): This vehicle followed the Ausf F from May 1940 and was in most respects identical. It was intended to have the new 5 cm gun, but about the first third of the production run still had the 3.7 cm gun. There was an improved cupola with five vision ports which had narrow twin sliding covers and thicker armour. There was also a new driver's visor with a single cover intended to give better splash protection. Vehicles sent to North Africa in 1941 were fitted with extra air filters and a faster working cooling fan and were known as **PzKpfw III Ausf G (Trop)**; some Ausf F vehicles were retrospectively altered to this standard to equip the Afrika Korps when the Germans became involved in the desert fighting from April 1941. Some 450 Ausf G vehicles were built. Chassis numbers: 65001 - 66000. Ausf F and G were identical in construction except that the turret of Ausf G was amended to accept either the old or new cupola.

Typ 7/ZW/PzKpfw III Ausf H(Sd Kfz 141): Major changes were made in this late 1940 production model in the light of the extensive combat experience from France which was now available. The track was increased in width (from 360 mm to 400 mm); there was a new final drive, the return rollers were spaced further apart; there was a new sprocket with six D-shape holes and an eight-spoke open idler wheel. Finally the suspension bars were strengthened, the whole effect of these changes being to give an improved ride, especially in desert terrain. A simple mechanical gearbox was also introduced, a Maybach SSG 77 all-syncromesh unit with six forward speeds and reverse. This lacked a pre-selector. Changes were made in armour protection with extra 30 mm plates bolted on to both upper and lower nose plates. A new turret with single piece rear plate was introduced on this model. Spaced armour was also retrospectively added on the superstructure face in some refitted vehicles, though was more common in later models. Some earlier models were subsequently refitted to roughly Ausf H standard, complete with new tracks and improved suspension. With its extra armour the Ausf H increased its weight to 21.6 tons. Chassis numbers: 66001 - 68000.

VK 2001(DB)/ Typ ZW40: This was a prototype produced during the production period of the PzKpfw III, intended as an improved version of the original PzKpfw III design. It was not connected with the 1935 VK 2000 series projects except that it revived the interleaved suspension and was generally based on the old specifications. Two prototypes were built and tested in 1939-40 by Daimler-Benz. The first prototype had 3-4 speed transmission, and the second a DB clutch and brake steering system. The engine was a Daimler-Benz MB 803 12 cylinder diesel unit. The vehicle weighed 22 tons and had the same armament and crew as the standard production PzKpfw III of the period. The suspension was of the FAMO type, fully-interleaved, as in contemporary German half-tracks. Top speed was 50 kph. The hulls and superstructures of the prototypes were taken from the PzKpfw III Ausf G production run, with the turrets the same as those for the Ausf H. Before any official assessment could be made the Russian campaign had star-

TOP: PzKpfw III Ausf H with spaced armour. CENTRE: ZW 40 in service for training (IWM-STT4064). ABOVE: VK 2001 (DB)/ZW 40.

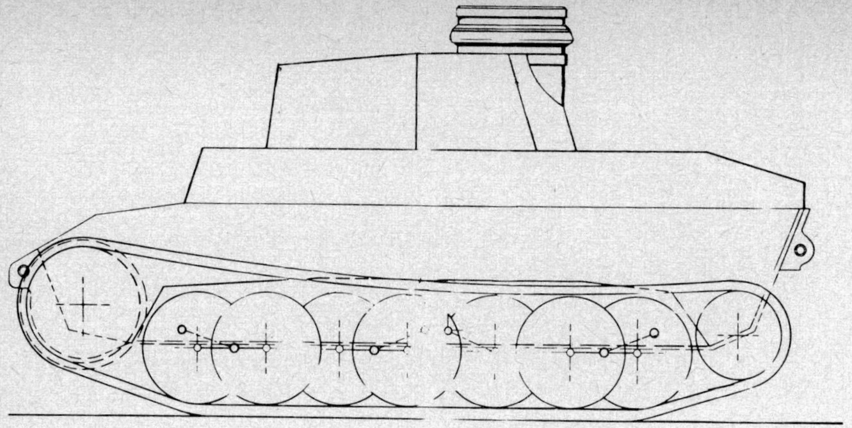

VK. 2001(D).

ted and the need was for rapid production of existing models. The ZW40 prototypes were therefore not adopted as production models. This ZW40 prototype has been described in the past as the PzKpfw III/IV, but no documentary evidence of this has appeared at the time of writing.

Typ 8/ZW/PzKpfw III Ausf J (Sd Kfz 141/1): Hitler suggested the fitting of a superior 5 cm weapon, the 5 cm KwK 39 L/60 'long' gun and ordered that it should become standard for the PzKpfw III in a directive made in August 1940. At a birthday demonstration of new tanks in April 1941, Hitler noticed that the 'new' PzKpfw III on display, which was the Ausf J, still retained the L/42 gun, and was most displeased to find that his orders had apparently been disregarded. However, in July 1940, just before Hitler's August directive, only 17 L/42 guns were ready, and only 84 of them were ready by the end of August. It is easy to understand, therefore, the reluctance of Heereswaffenamt to switch immediately to the L/60 gun even before the L/42 had been tried in action, and there was, in any case, an inevitable delay in setting up new production.

Production of the Ausf J started in March 1941, and important changes included increase of basic armour from 30 mm to 50 mm at the front and 32 mm at the sides. Because of the increased frontal thickness the hull machine gun was placed in a ball mount and the driver's visor was revised. The first production series (chassis numbers: 68001-69000) all had the 5 cm L/42 gun. The 'long' L/60 gun was introduced in the second series (chassis numbers: 72001-74100). Even so about 500 of these must have been built with the L/42 gun though the first vehicles with the L/60 gun appeared in December (or possibly November) 1941. Vehicles with the L/42 and L/60 guns were produced concurrently during the changeover period. On the second series (from vehicle 72001) spaced armour was standard on mantlet and superstructure fronts, and vision slots were accordingly eliminated from the mantlet. The spaced armour gave protection from the hollow charge type rounds then coming into use with anti-tank weapons. With the L/60 gun, this vehicle

TOP: PzKpfw III Ausf J. CENTRE: PzKpfw III Ausf H, front detail. ABOVE: PzKpfw III Ausf L (IWM-STT3969).

became known as the 'Mk 3 Special' to the British and was very successful in the Western Desert in giving the German panzer troops superiority over British tanks. Vehicles with the long gun had the ammunition stowage reduced from 99 to 84 rounds.

Typ 9/ZW / PzKpfw III Ausf L (Sd Kfz 141/1): In this model, built from July 1942, the L/60 gun was fitted as standard from the start. As the long gun tended to make the vehicle nose heavy, a coil ring compensator, and later a torsion bar, were fitted in the gun mount. The suspension was unchanged except that the diameter of the forward torsion bars was increased to 55 mm (first three), the others being 44 mm. Spaced armour was again standard on mantlet and superstructure fronts. The spaced armour consisted of 20 mm plates supported on a metal frame attached to the front superstructure and mantlet. Holes were cut in the spaced armour plates to clear the machine guns, main gun, and driver's visor. The spaced armour was intended to counteract the effects of hollow charge ammunition at points of special vulnerability. The rear smoke emitters were moved to a central position under the engine tail plate. On all but the earliest vehicles the various superstructure and turret side vision ports were dispensed with to simplify production. A major sub-variant was designated **PzKpfw III Ausf L(Tp),** a tropicalised version for the desert theatre which saw wide service with the Afrika Korps in 1942-43. Changes included larger oil filters, larger single engine cover doors replacing double doors. Larger fan access doors were fitted and all engine cover doors had air intake vents. Air was trunked into the fighting compartment via the engine compartment intakes to improve habitability. A curved metal sheet was fitted under the rear air outlet so as to deflect the air flow from the moving vehicle and so reduce the size of the dust cloud. Ammunition stowage was reduced to 78 rounds. Chassis numbers: 74101-76000.

Half the vehicles of this order were fitted with the short 7.5 cm gun, like the Ausf N, as the PzKpfw IV with 7.5 cm L/43 had become the principal battle tank by the time the PzKpfw III Ausf L was in production.

Typ 10/ZW/PzKpfw III Ausf M(Sd Kfz 141/1): This was essentially a simplified and updated Ausf L with detail changes, built from late 1942. All pistol ports, vision ports, and the hull side escape hatches were eliminated. The tail smoke emitters were deleted and triple smoke dischargers (90 mm NbKwg) were fitted each side of the turret firing forward. Spaced armour was standard on superstructure front and mantlet. The later production vehicles had full skirt armour added to the turret and hull sides to offer added protection and some earlier vehicles were also refitted to this standard. The interior ammunition stowage was changed to increase capacity slightly. Because of all the extra armour a coil spring was added to the front suspension units. A deep wading kit was added to these vehicles allowing them to wade up to 4 ft 3 ins. Seals were added to the superstructure doors, at the hull and superstructure join, and in the engine exhaust outlets. There was also a coolant transfer system. Chassis numbers: 76001-78000.

Typ 11/ZW/PzKpfw III Ausf N (Sd Kfz 141/2): The last of the PzKpfw III series in combat tank form was in production from January 1943, building starting at the tail end of 1942. These all had the short KwK L/24 7.5 cm gun.

ABOVE: PzKpfw III Ausf L, rear detail. LEFT: PzKpfw III Ausf M—note wading gear and NbKwg on turret. BELOW: PzKpfw III Ausf L completed with 7.5 cm L/24 gun. PzKpfw III Ausf N was of similar appearance (Axel Duckert).

The N was similar to the M in detail, but had a new cupola (as fitted to the PzKpfw IV) with 100 mm armour, and the hinges deleted from the brake access hatches. This vehicle was also known as the Sturmpanzer III. Ammunition capacity was increased to 64 rounds of 7.5 cm ammunition and 3450 rounds of machine gun ammunition. Chassis numbers: 76001–78000.

By the time the Ausf M appeared the PzKpfw III had reached the limit of its development potential as a tank, due to the fact that it could not carry the long 7.5 cm gun (it had a 128 cm diameter turret ring) which was necessary in the 1942-43 period to combat the latest Allied tanks like the Soviet T-34 and the American M4 Medium(Sherman). Ironically enough, the bigger PzKpfw IV, was able to absorb upgunning (it had a 152 cm diameter turret ring) and so became the principal combat tank, taking over the role envisaged pre-war for the PzKpfw III. In the Ausf M and N form, with the short low velocity 7.5 cm gun originally carried in the PzKpfw IV, the last model of the PzKpfw III series therefore took over the 'close support' role (Sicherheitspanzer) originally envisaged for the PzKpfw IV. In August 1943 PzKpfw III tank production was terminated. It had effectively ended in December 1942, but sporadic production (43 in May, 11 in June, 20 in August) continued into 1943 to use up remaining components. Henschel, Daimler-Benz and MAN all switched their production lines to Panther tanks, and Alkett and MIAG switched over to producing the StuG III which had a chassis derived from the PzKpfw III. In late 1942 and early 1943 some actual PzKpfw III chassis were used to increase StuG III production.

PzKpfw III Production

When the ZW was designed and first produced mass-production on the scale ultimately needed was not foreseen. However, by 1939 the urgency of the situation, with tanks still in short supply, led to contracts being given to many firms who formed a PzKpfw III production group. Alkett, Daimler-Benz, FAMO, Henschel, MAN, MIAG, Wegmann, and MNH were all involved in manufacture or assembly or both. Major sub-assemblies were made by Deutsche Edelstahlwerke, while Karges-Hammer and Garny were firms making the guns. Though mass-production was not originally envisaged, the design allowed for it, with the vehicle being broken down into four major sub-assemblies, hull, turret, front superstructure, and rear superstructure. All structures were single skin and welded, and the sub-assemblies were bolted together at final assembly. A production target of 108 vehicles a month was fixed in November 1940 but by July 1941, after the invasion of Russia and the decision to expand the armoured forces to 36 Panzer Divisions, a requirement of 7,992 vehicles was set. Targets were rarely met, however, and the overall production total for all models of PzKpfw III was 5,644.

The PzKpfw III was quite a straightforward vehicle. The engine was located in the rear and early models to G had flush engine cover doors; thereafter raised vented doors were fitted, and also appeared on refitted early models. The gearbox and steering mechanism was in the forward hull compartment with access doors in the nose. The driver and radio-operator/hull gunner sat in the front of the fighting compartment. The driver had a vision block in the superstructure front. The hull gun was fitted in a Kugelblende 30

TOP: PzKpfw III Ausf F Unterwassen Tauchpanzer on test, 1940 (BA-MW5674-33).
ABOVE: PzKpfw III Ausf M(Flamm) in action (BA-294-1536-35).

gimbal mount in models A-H, and thereafter in a Kugelblende 50 ball mount. The former was distinguished by an external square-shaped armoured frame. Side visors with vision blocks were fitted for the two forward crew members, and there were side escape doors low in the hull in all models up to L; thereafter these side openings were deleted to simplify production. The remaining three crewmen were in the turret; the turret had no cage but the commander and gunner had seats attached to the turret walls. Hinged doors, vision and pistol ports were fitted in the turret sides, the latter being deleted from later models. Drive from the engine was taken via a cardan shaft forward through the fighting compartment under a false floor to the steering and transmission compartment in the nose. Here was situated the gearbox and the Daimler-Benz/Wilson hydraulic clutch braking system. Many of the components of the PzKpfw III were interchangeable with the PzKpfw IV.

The engine compartment was separated from the fighting compartment by a bulkhead. The engine was set in the centre with a fuel tank and battery box on either side. Two radiators were placed at the rear of the engine.

The turret was traversed by hand by the gunner, but an auxiliary traverse was provided for the loader which enabled the turret to be moved more quickly when both were used. The commander and gunner had turret and target indicators. The radio fitted in the tank with the Fu5, standard for all German tanks of the World War 2 period. The external radio aerial, mounted on the right side, was spring loaded and could be deflected by overhanging obstacles, returning to its vertical position, or it could be locked back.

Special Purpose Variants

Aside from the basic production models, there were many adaptations or conversions of the PzKpfw III for special purposes, and these are outlined here.

PzKpfw III Ausf F/G/H Unterwassen Tauchpanzer: Special armoured battalions were formed to train in the technique of landing tanks from the sea, to take part in the proposed invasion of Britain (Seelowe). The PzKpfw II was light enough to be given flotation chambers for 'swimming' ashore, but the PzKpfw III and IV were considered too heavy for this and these vehicles were modified so that they could operate by driving ashore from a barge by way of the sea bed (Unterwassen Tauchpanzer = under water submersing tank). The tank was made entirely watertight by the use of sealing compounds on all opening parts. The turret hatches were secured from the inside, and the engine air intakes were closed. Rubber covers were fitted over the gun muzzles, hull machine gun, and cupola, with a charge to blow them off on emerging from the water. The turret and hull join was sealed by an inflatable rubber tube. A 50 ft long flexible hose Snorkel tube, buoyed at the top end, provided an air supply. A radio aerial was attached to the buoy. A bilge pump in the tank removed any water which seeped through the waterproofing. On approaching the landing area, the ramp of the landing barge was lowered and was long enough to reach the sea bed at a depth of roughly 25 ft. A gyro-compass aided underwater steering, plus radio commands from an observer aboard ship. Current production vehicles of the Ausf F, G and H types were converted for this role. Seelowe was postponed, and the vehicles which had

TOP: PzKpfw III Ausf M(Flamm). Note spaced front armour. CENTRE: Minenraumpanzer III, rear view. ABOVE: Pioneerpanzerwagen III with skirt armour.

been prepared were then used for the River Bug crossing in 1941 during the invasion of Russia. The Snorkel was replaced by a 10 ft steel tube and one-way valves were fitted to the exhaust.

PzKpfw III Ausf M(F1)(Flammpanzer III)(Sd Kfz 141/3): In early 1943 the firm of Wegmann was given a contract to convert 100 new PzKpfw III Ausf M to flame-throwing tanks (Flammpanzer). The flame-throwing unit was worked by pump, the projector being within a 5 ft long steel tube which superficially resembled the long 5 cm gun. The normal machine gun mounts were installed. Some 225 gallons of flame fuel was installed in two tanks inside the hull. The projector could fire 70-80 flame shots of 2-3 seconds duration. Range was 65 yards maximum and 40 yards affective. A three-man crew was carried, extra armour was welded on the nose, and the weight was increased to 23 tons. An extra radio set (Fu 2) was installed. Most were used on the Russian front.

Minenraumpanzer III: This was a special conversion with raised hull and strengthened and much modified suspension. It was based on the Ausf F. The vehicle was used to propel a mine-clearing device and a number are known to have been converted.

Pioneerpanzerwagen III: This was a conversion of a PzKpfw III by removal of the turret and with wooden decking added. It was used to carry bridging materials and engineers' stores. A good number of these were produced after the PzKpfw III became obsolete.

Munitionspanzer III: A similar conversion to the Pioneerpanzerwagen, this had the turret removed and was used to carry ammunition, with suitable racks being arranged in the gutted fighting compartment.

Schlepper III: This was a PzKpfw III with turret removed used for general towing and store-carrying work. A wood platform was built on the superstructure to carry crated stores in those vehicles used on the Eastern Front later in the war. Most vehicles used in this theatre had Ostketten tracks (grousers) fitted for improved performance in snow and mud. Numbers were converted after the PzKpfw III became obsolete. However, even in the early years of the war there were a number of turretless Schleppers in service, particularly at factories and ordnance depots where they were used to tow 'dead' vehicles.

Bergepanzer III: This was a turretless PzKpfw III converted as an armoured recovery vehicle. Suitable tools and stores were carried for the recovery role, but the vehicle had only a towing capability, not a lifting ability. In 1944 when many obsolete chassis were available, over 100 were converted.

Eisenbahn Wagen PzKpfw III: A small number of PzKpfw IIIs were converted in 1943 to run on rail tracks and were used on the Eastern Front to protect the long stretches of railway track.

Grosser Panzerbefehlswagen III/Gr PzBefWg III(Sd Kfz 266, Sd Kfz 267, Sd Kfz 268): A series of command vehicles was produced throughout the war period, by adaptation from the basic PzKpfw III. The first batch (chassis numbers: 60341-70) were designated Ausf D1 (Typ 3c/ZW), and were built from mid 1938 to 1939 concurrent with the standard Ausf D tank. The next batch, designated Ausf E (chassis numbers: 60501-45), were

TOP: Bergepanzer III recovering early model PzKpfw III; Schlepper III and Munitionspanzer III similar to Bergepanzer III. CENTRE: Eisenbahn Wagen PzKpfw III Ausf N (Werner Oswald). ABOVE: Gr PzBefWg III Ausf H.

produced from mid 1939 to 1940; though based on the Ausf E tank some late vehicles had details similar to the Ausf G(eg, the driver's visor). The Ausf H (chassis numbers: 70001-175) was theoretically based on the Ausf H tank, originally with dummy 3.7 cm gun, but later vehicles were actually completed with the fittings and characteristics of the Ausf J, late model, complete with 5 cm L/42 gun, and spaced armour. Next came Ausf K, built late 1942 to 1943, which had the normal 5 cm L/60 gun, extra visors, and the final type of cupola with 100 mm armour. It was based on the Ausf M tank. Finally there were a large number of Umbau PzBfWg III Ausf L (umbau = converted), around two-thirds of the output of this model being eventually converted to command vehicles when the PzKpfw III went out of divisional service from 1943. PzBfWg III Ausf D1 and E had the turrets fixed in position and bolted to the superstructure. A dummy gun resembling the 3.7 cm weapon was fixed to the mantlet and a ball-mounted machine gun in the mantlet was the actual main armament, for defensive use only. A pistol port replaced the hull machine gun. A prominent rail type aerial was carried round the rear hull decking, and there were two rod aerials and a star aerial on a 9 metre winch mast. On the subsequent models the normal mast type of aerial replaced the rail and star aerials. The Ausf K resulted from a need for a tank which could be quickly put into service as a command vehicle when required – it enabled a commander to transfer from vehicle to vehicle. Provision was made to accommodate the Fu 8 radio set or any other set as required on a portable basis, the sets being shifted to suit needs. The loader acted as an extra radio operator. Map tables and binocular observation periscopes (forward of the cupola) were fitted in these command vehicle models.

The ordnance designations indicated the particular radio equipment fitted

TOP: Gr PzBefWg III Ausf D1—note Ausf D chassis. CENTRE: Late Production Gr PzBefWg III Ausf H (RAC-243-38). LEFT: Turret detail, early Gr Pz PzBefWg III Ausf H. OPPOSITE PAGE: Gr PzBefWg III Ausf H, very late vehicle with Ausf J fittings.

in the vehicle, as follows:

Sd Kfz 266: Fu 13 (battalion or regimental link)

Sd Kfz 267: Fu 13, Fu 6, Fu 8. (main divisional link).

Sd Kfz 268: Fu 13, Fu 6, Fu 8, Fu 7. (ground-to-air link).

All vehicles could operate on the tank frequency band.

Panzerbeobachtungswagen III/ PzBeobWg III(Sd Kfz 143): In 1943 a special armoured observation vehicle was converted from older PzKpfw IIIs in for refitting. Main vehicles were Ausf E, F, and G. A dummy gun barrel was fitted, offset from the centre-line. There was a ball-mounted machine gun in the centre of the mantlet. The vehicle was intended mainly for control work with self-propelled gun batteries. Radio fitted included Fu 8 and Fu 4, and a loudspeaker. Observation gear included periscope binoculars. There was a crew of four. Later in the war some PzBeobWg IIIs were converted from late model tanks of the L, M, and N types, which retained their main armament.

Pz Kpfw III(Pak 7.62 cm (r)): A few PzKpfw IIIs captured by the Soviets were re-armed by the Russians with the standard 7.62 cm tank gun (as fitted in the T-34). The Red Army designated this conversion SU 76i. Some were re-captured by the Germans who put them back into service with the appropriate designation. As converted, they became in effect true assault guns, for the armament had only limited traverse within a fixed superstructure the full width of the hull and open at the top. Frontal armour was 30-40 mm and side armour 10-15 mm on the new superstructure.

BELOW: PzBeobWg III (W.J. Spielberger). OPPOSITE, TOP TO BOTTOM: Gr PzBefWg III Ausf K. Umbau Gr PzBefWg III Ausf L with skirt armour and AA MG mount on cupola. PzKpfw III (Pak 7.62 cm (r)).

7: Panzerkampfwagen IV

THOUGH a prototype VK 2001 (Rh-B), for the BW requirement was completed in mild steel by Rheinmetall-Borsig in late 1934, and tested through 1935-36, the development contract for this 'medium' or 'close support' tank finally went to Krupp in 1935. This firm already had the contract to develop the turret and main armament for the tank. The Krupp design made use of the features which they had presented (unsuccessfully) for the ZW 'light' tank requirement. Krupp's MKA prototype already had the hull shape and form, and the simple suspension which was to become a hallmark of the PzKpfw IV. Chronologically, the PzKpfw IV preceded the PzKpfw III into service, and outlived it by a wide margin. The PzKpfw IV was to become the most important German tank numerically to see service in World War 2, and it was the only one to be in production throughout the entire World War 2 period. The basic design was proved sound and reliable, and though it was considerably updated to keep pace with changing combat and technological requirements, there was no radical change in design throughout its production life save for upgunning and uparmouring.

Though the PzKpfw IV/BW was slow to come into production, and was only built in small pre-production batches to start with, the design made provision for mass-production. The vehicle was divided into four sub-assemblies, consisting of the hull, front superstructure, rear superstructure, and turret. All were complete structural units which bolted together in final assembly. The hull itself was divided into three, by two bulkheads. The engine was mounted at the rear and the drive shaft was taken forward under the fighting compartment floor to the clutch and gearbox in the driving compartment. A DKW 500 cc auxiliary two-stroke petrol engine was provided to drive the generator used for the turret's power traverse, sited to the left of the main engine but removed in late production vehicles in favour of extra fuel storage. Three fuel tanks were sited under the floor beneath the turret turntable. The rear superstructure could be removed to facilitate engine changing. The superstructure overhanging the hull sides, allowing a generous (168 cm) diameter turret ring and good internal stowage. The turret was well sloped and the cage consisted of four tubular members connecting the turret to the turntable floor. The 7.5 cm gun, in either its short or, later, long forms, was mounted on the centreline with its trunnion axis just behind the rear of the turret front face.

Driver and radio operator were on the left and right sides respectively of the front compartment with the gearbox between them. Each has a roof hatch, and the driver had a vision flap and a binocular episcope. The hull machine gun was ball-mounted in the right front superstructure. Hinged flaps on the glacis plate gave access to the steering mechanism in the nose. In the fighting compartment the turret traverse gear and gun elevating gear was on the left side of the gun all operated by the gunner. He had a sighting telescope or the option of an open sight bar in the left front port of the turret. The gun-

TOP: Wood mock-up of BW by Krupp. ABOVE: PzKpfw IV Ausf A, showing turret rear. BELOW: VsKfz 622 pilot model.

ner fired the gun by an electrical trigger on the traversing handwheel and used a foot pedal to fire the coaxial machine gun (MG 34). The commander sat immediately behind the deflector plate of the gun, provided with an armoured cupola towards the rear centreline of the turret. There was a tip-up seat provided, and the commander could stand on the turret floor, sit in the cupola, or stand on a footrest looking out of the cupola top. The loader had a tip-up seat to the right of the gun, but normally stood to load the 7.5 gun and the co-axial machine gun, and ensured an adequate supply of ready-use ammunition. There were vision ports in the front of the turret and on each side. On each side of the turret was an access door, also with an armoured vision port, and pistol ports were provided also, depending on the model.

There were four bogies each side, each carrying two rubber-tyred wheels. Quarter elliptic leaf springs were anchored under the leading axle of each bogie with the tail of the spring resting on a roller under the trailing axle. There were four return rollers and a front drive sprocket and rear idler on each side. Rear smoke emitters were fitted on the hull rear of early models, and these were replaced in later models by smoke discharger cups on the turret sides, fired electrically. The radio aerial was spring loaded and could be deflected into its trough by a special rail on the gun assembly (or the gun barrel itself in later models). Or the 2 metre aerial could be retracted inside the tank by the crew. Other details varied from model to model and these are described in production order.

VsKfz 622/Typ 1/BW/PzKpfw IV Ausf A: The designation Versuchskraftfahrzeug 622 (Experimental Vehicle 622) was given to the first Batch of BWs built to the Krupp design. These numbered 35 only and were used for trials and training. The engine was a Maybach 250 HP HL 108 TR. Cooling was achieved by taking air through the right side grilles, passing it through the radiator, and expelling it through the left side grilles. There was a five-speed ZF SFG 75 gearbox giving a top speed of 30 kph. Transmission was via a dry-plate three-disc clutch and a 5/6 speed gearing to the sprocket wheel. Steering was by Wilson-Krupp type clutch and brake. The 7.5 cm KwK L/24 was then newly developed and was in an internal mantlet. The superstructure was shaped in front to place the driver slightly further forward than the radio operator. The commander's cupola was a simple drum-shaped unit built into the back plate of the turret. Circular signal ports (two) were in the turret roof forward of the cupola. There was a single D-shaped access door each side of the turret. These vehicles were built in 1937-39. Chassis numbers: 80101-135.

Typ 2/BW/ PzKpfw IV Ausf B: Some 45 of these were built in 1937 and these were in service during the Poland campaign in 1939. Among changes made were a more powerful engine, the Maybach HL 120 TR of 320 HP, and a new six-speed syncromesh gearbox, the SSG 76. The frontal armour was increased from 14.5 mm on the Ausf A to 30 mm on the Ausf B. This took the battle weight from 17.1 to 17.5 tons. The front of the superstructure was changed to give a one-piece straight unit. The driver's visor was changed from a single flap to twin sliding shutters, and pistol port replaced the bell mounted hull machine gun and a vision slit was also provided in the front plate. Single access doors replaced the twin doors which were provided for the two front crew members in the Ausf A. An armoured cupola replaced the simple drum

TOP: PzKpfw IV Ausf B or C. CENTRE: PzKpfw IV Ausf B or C. ABOVE: PzKpfw IV Ausf D.

type, with five vision ports covered by armoured sliding shutters. Square turret pistol ports were replaced by oval ones, and armoured shutters were provided over the mantlet vision ports. The left hand signal port was covered by a cone-shaped hood (Haubecap) which contained a signal lamp. Chassis numbers: 80201-242.

Typ 3/BW/PzKpfw IV Ausf C: This was built in 1938-39 and was an important model, being the main type in service for the invasion of Poland. It was identical to the Ausf B except for an increase of armour from 20 mm to 30 mm in the turret front, and an armoured sleeve over the co-axial machine gun. From the 40th vehicle on a new engine was installed, the Maybach HL TRM of 300 HP, which incorporated an impulse magneto. Chassis numbers: 80301-440.

Typ 4/BW and Typ 5/BWPzKpfw IV Ausf D (Sd Kfz 161): The Model D appeared in September 1939 and had several detail improvements. The internal mantlet gave way to an external version with a curved 35 mm armour shield, this extending the full width of the mantlet. Turret and cupola were unchanged otherwise, except for minor changes to the vision slots on the turret sides. The front superstructure was now changed back to the original pattern similar to the Ausf A with the driver forward of the radio operator's station. A pistol port was provided in the forward extension of the superstructure. The hull machine gun was returned to its original position, now with a square-frame gimbal type mounting, the Kugelblende 30, as in the PzKpfw III. Double shutter driver's visor was retained. Spare track shoes were usually secured across the nose and glacis for extra protection. Vehicles returned for refit from late 1940 were given a new engine cover, and a stowage bin was fitted on the turret rear. A new drive sprocket was added together with wider tacks having 99 rather than 101 track shoes. Limit stops were fitted to all the bogies rather than to just the front and rear units as originally designed. In August 1942, Hitler directed that all PzKpfw IV early models returned to the factory for repair were to be fitted with the long barrel 7.5 cm KwK 40 L/43 to bring them up to Ausf F2 and G standards. The upgunned vehicles had the turret vent and the signal ports plated over and a fan vent fitted. Armoured skirting was fitted permamently round the turret. Extra armour was provided by bolting 30 mm plates on the superstructure front and 20 mm plates on the sides back to the engine compartment. Others were put on the fighting compartment sides. While this was originally a field modification it became a standard factory practice on refits too. In actual fact, however, not all the early models were so extensively refitted. Chassis numbers: 80501-748.

Typ 6/BW/PzKpfw IV Ausf E(Sd Kfz 161): In late 1940 a new model went into production with several changes. It was contemporary to the uparmoured PzKpfw III Ausf H, and saw action against the British in North Africa in the summer of 1941. A new thicker cupola was fitted, having narrow sliding visors, and the turret rear angle was changed so that a single curved plate could be used without the need of a cut-away to clear the cupola. On the turret roof the right signal port and the central roof vent were eliminated, and a circular fan vent was installed forward and to the right. The vent cowl was protected by a circular guard and there was an electric ex-

TOP: Turret of PzKpfw IV Ausf B/C. ABOVE: Uparmoured and refitted Ausf D, with wide tracks. BELOW: PzKpfw IV Ausf E.

tractor fan fitted beneath the vent. The superstructure front was like the Ausf D but a new single hinged driver's visor, known as a Fahrershklappe 30, replaced the previous type. The most important change was that the nose plate now became a single thickness of 50 mm armour, but the extra bolted armour plate was continued on the sides and the driver's plate, though not all vehicles had this; this work was done in the factory. A new type of driving sprocket was now used, of simpler design and allowing the use of wider (400 mm) track. There were new pattern brake access hatches on the glacis, with a single big hinge replacing two small hinges. Chassis numbers: 80801 - 81023.

Typ 7/BW/PzKpfw IV Ausf F1(Sd Kfz 161): Appearing in 1941 at about the same time as the PzKpfw III Ausf J, this model saw substantial changes. The basic armour thickness was increased from 30 mm to 50 mm on all front faces (hull, superstructure, turret) and to 30 mm on the sides. The supplementary bolted armour was dropped. The superstructure front was changed back to a single straight plate, and the hull machine gun was changed from the gimbal mount to a ball mount, Kugelblende 50. A thicker driver's visor, Fahrersehklappe 50, was also incorporated. The large single doors on the turret sides were replaced by double doors which had a vision port in the leading door and a pistol port in the rear door. Air intakes with cast cowls were included on the glacis plate brake access hatches. The new pattern dished sprocket (also on the late Ausf E) and a tubular construction idler wheel were fitted on each side, taking the new 400 mm wide track which included slots in the outer web to take grousers or ice sprags. With the invasion of Russia and the requirement for many more Panzer Divisions, PzKpfw IV production was greatly stepped up. Two more firms, Vomag and Nibelungenwerke subsequently joined Krupp in building this model. Chassis numbers: 82001 - 396 (Krupp), 82501 - 575 (Vomag), 82601 - 614 (Nibelungenwerke).

Typ 7/BW/PzKpfw IV Ausf F2(Sd Kfz 161): The tail end of Ausf F production was now converted to take the long barrel 7.5 cm KwK 40 L/43 gun on Hitler's direct orders, consequent upon the situation on the Russian Front where the T-34 had appeared and out-gunned all existing German tanks. The long barrel gun had been planned for introduction in the Ausf G, but events dictated its priority fitting in the earlier model. Fitted with the long 7.5 cm gun the PzKpfw IV became a reasonable match for the T-34, and superior to British tanks. Vehicles so fitted were popularly known to the British as the 'Mk IV Special'. Aside from the gun there were no other factory changes in the Ausf F2, though subsequent refits could result in installation of fittings more common to later models.

Typ 7/BW/PzKpw IV Ausf G(Sd Kfz 161/1): This vehicle became the most important of the PzKpfw IV series, being built in huge numbers and with many production changes. The BW designation was dropped from this model on. The early KwK 40 gun had a single baffle muzzle brake and was to be seen on both the F2 and G. A double baffle muzzle brake was soon introduced and became the commonest. On some early Ausf G the vision ports in the turret sides and front were deleted for both gunner and loader, but generally only the loader's were so removed. The turret roof was thickened from 10 mm to 15 mm and there were many detail changes internally. There were very many detail changes externally — Ausf G tanks with Ausf F2

ABOVE: Early PzKpfw IV Ausf F2 with single-baffle muzzle brake.

ABOVE: PzKpfw IV Ausf Fl. RIGHT: PzkPfw IV Ausf F2.

turrets, late production vehicles with early KwK 40 guns, and so on. Ausf G production was under way by mid 1942. From late 1942 an improved Ausf G appeared, production changes including the longer L/48 gun replacing the L/43. From July 1942 a proportion of tanks produced had extra 30 mm armour plates welded or bolted to the nose and front. By January 1943 all production vehicles had the additional armour. At this time a new cupola with 100 mm armour and single piece hatch appeared, and finally hull and turret side skirting plates (schurzen) were introduced. Chassis numbers ran to 84000. The three manufacturers, Nibelungenwerke, Krupp, and Vomag were each allocated separate number bands, but within these bands not all numbers were necessarily taken up by production.

Typ 7/BW/PzKpfw IV Ausf H(Sd Kfz 161/2): This was a new production model which introduced several mechanical and constructional features when introduced in mid 1943. The 7.5 cm KwK L/48, first introduced on the late Ausf G was standard, as was the thickened (100 mm) cupola, which had a single flap rather than two half flaps. On some vehicles a V-shaped splash plate was welded to the turret roof forward of the cupola. Like the late Ausf G, the Ausf H had extra 30 mm armour plates welded or bolted to the nose and front vertical plates. A new eight-spoked driving sprocket of webbed pattern was fitted, and the gearbox was changed from a ZF SSG 76 to an improved type, the ZF SSG 77. The very late production Ausf H had entirely new armour disposition, a single thickness of 85 mm replacing the welded or bolted layers, and the armour plates were of inter-locked construction. Very late vehicles also had a new idler wheel of webbed and welded construction, the return rollers were of all steel type, and simplified limiting stops were fitted for the bogies. New vehicles were all fitted with Schurzen as standard, these being intended mainly as protection against hollow-charge projectiles of the bazooka type. The side skirts were mild steel, 5-9 mm thick, hung from rails attached to the vehicle sides. These plates were removable. The turret skirting was similar but was a permanent fixture, with hinged panels opposite the turret side doors. Zimmerit anti-magnetic mine paste was a usual covering on new vehicles in the 1943-44 period. A semi-circular rail, Fliegerbesschussgerat, was fitted to the front of the cupola in most vehicles to take a MG 34 mount for AA defence. Output of the Ausf H was huge, Nibelungenwerke being the major producer with 46% of production. Krupp and Vomag built 27% each. Chassis numbers: 84001 - 90000 in at least six production bands.

PzKpfw IV Ausf J(Sd Kfz 161/2): In December 1943 Krupp had ceased PzKpfw IV production (switching to StuG IV and the Mobelwagen FlPz IV), and in mid 1944 Vomag switched to building the Jagdpanzer IV. This left only Nibelungenwerke in Austria to build the last model, the PzKpfw IV Ausf J which remained in production until the war's end. The Ausf J was similar to the Ausf H but featured a number of changes to simplify production. The auxiliary two-stroke DKW 500cc generator was removed from the engine compartment. An extra 44 gallon fuel tank replaced it. The turret could now be traversed by hand only. All turret vision ports were deleted except the gunners. The skirt armour was retained, but on late models the hull plates were replaced by heavy gauge wire mesh. A wider track specially

TOP: PzKpfw IV Ausf F2. ABOVE: PzKpfw IV Ausf G—early production. RIGHT: PzKpfw IV Ausf G—late production with skirt armour.

developed for use in ice and snow on the Russian Front (Oskette = East track) could be fitted. On late vehicles the four return rollers were reduced to three. A new type of turret fan extractor was fitted, and the rear exhaust system was simplified. Many fittings were now common to both PzKpfw III and IV and were like those used on SP vehicles based on the unified Pz III/IV chassis. For example towing lugs were formed from an extension of the side plates, the same as those introduced on the PzKpfw III Ausf J.

Panzer IV/70(A)(Sd Kfz 162/1): It was planned to end PzKpfw IV production by February 1945, but production continued to the end of the war, due to the diminishing speed of production of the order. However, the success of the Vomag-designed Jagdpanzer IV led Hitler to ask for the premature end of PzKpfw IV production in July 1944. He wanted all PzKpfw IV output to be converted to Jagdpanzer vehicles armed with the 7.5 cm L/70. It was not possible to consider immediately such a change as production was then about 300 units a month with considerable advance orders placed with suppliers. It was agreed that Nibelungenwerke would convert to the new Jagdpanzer IV chassis as soon as possible but in the meantime a proportion of their PzKpfw IV production would be converted to the Pz IV/70 Zwischenlosung (= interim) with a superstructure designed by Alkett. Thus by the end of the war, of the 2,500 (approximately) Ausf J built by Nibelungenwerke, almost 300 were converted to Pz IV/70(A) or Zwischenlosung. This vehicle was not a tank in the accepted sense, being really a version of the Jagdpanzer IV under another name. Its full development history is intimately bound up with the story of German self-propelled equipment and is outside the scope of this present book. The Pz IV/70(A) was, however, built on a tank chassis and justifies its inclusion here. There were several detail variations in the chassis. Most were essentially the Ausf J chassis, but some Pz I V/70 were on the unified PzKpfw III/IV chassis. The two leading bogies each side had steel rimmed 'silentbloc' wheels, the rear bogies having rubber tyred wheels.

PzBefWg IV (7.5 cm KwK L/48): This conversion to a command vehicle was carried out from early 1944 after the supply of PzBefWg III was K or Umbau L. About 100 were converted.

PzBeob Wg IV: A few new PzKpfw IV Ausf J were converted to armoured OP vehicles in the latter half of 1944 when the supply of PzBeobWg III was diminished. A pole type periscope alongside the cupola, and a radio aerial on the turret roof distinguished this vehicle which was externally similar in all other respects to the PzKpfw IV Ausf J.

Bergewagen IV: A few old PzKpfw IV chassis were used as towers, the turret being removed and recovery gear and tools being carried. A small number were converted late in 1944.

SP guns: The PzKpfw IV chassis was used as a basis for a number of self-propelled guns, including StuG IV, Brummbar, Jagdpanzer IV, FlPz IV Mobelwagen, Wirbelwind, and several other types.

Projects and prototypes: Several experimental conversions were made using the PzKpfw IV chassis. One vehicle was given hydraulic drive by the firm of Zahnradfabrik at Augsburg. The idler wheels were moved to the chassis front and the drive units were fitted at the rear under a sloped com-

TOP: Early production PzKpfw IV Ausf H. ABOVE: Late production PzKpfw IV Ausf H. LEFT: Turret and cupola detail Ausf H (IWM-BNA 7096).

bined engine and transmission cover. Test vehicle was an Ausf G. Another trials vehicle was fitted to operate from producer gas fuel, the fuel being carried externally in containers carried in a vertical box at the tail of the vehicle. Designs were drawn up for a PzKpfw IV model mounting the complete turret and L/70 gun from a Panther tank, (Schmal turm), but the prototype was never built due to the obvious overload on the chassis. A further project existed only in model form. This would have carried two recoilless rifles in a co-axial mount together with a 2 cm aiming gun.

Munitionspanzer IV 'Karl': This was a special conversion of the PzKpfw IV Ausf F with a load compartment and a 3 ton crane, specially designed to carry ammunition for the heavy mortar Karl. Only a token number were converted.

Bruckenleger IV: Two models of a launched vehicle-carrying bridge on the PzKpfw IV chassis are known to have been produced. One was built by Magirus and the other by Krupp. These were in service in 1940 and the Krupp-built vehicles were used in the invasion of France.

Bruckenleger IVS/Inf Sturm-Steg: This vehicle carried a light extending sectionalised bridge, total length 85 ft, enabling infantry to cross streams and gaps. The bridge was similar in structure to a fire brigade ladder.

PzKpfw IV Ausd D/E Unterwassen Tauchpanzer: Limited numbers of vehicles were converted for underwater operation in 1940 for the projected Operation Sealion. Features were similar to those described for the PzKpfw III conversion. In 1941 these vehicles were used in the invasion of Russia.

PzKpfw IV Ausf H converted for trials with producer gas fuel. Charcoal burners at rear.

TOP: Sturmsteg IV (assault bridgelayer). ABOVE: Bruckenleger IV (Magirus), showing method of using two or more bridges to span wide gaps. Note folding legs. BELOW: Bruckenleger IV (Krupp), rear view, France 1940.

TOP: PzKpfw IV Ausf J, late production—3 return rollers (C. Kliment). CENTRE: PzKpfw IV Ausf H, front detail. ABOVE: PzKpfw IV Ausf G with hydraulic drive, rear detail.

TOP: PzBeobWg IV; note periscope and late type cupola. ABOVE: PzKpfw IV Ausf D Unterwassen Tauchpanzer in Russia, 1941. Note rear exhaust valves. LEFT: Project model—PzKpfw IV with two recoilless rifles and 2 cm aiming gun.

TOP: Panzer IV/70 prototype. Production vehicles had one-piece sloped sides and most lacked skirt armour. CENTRE: PzKpfw IV Ausf D after 1942-43 rework, with long gun, turret armour, and extra front armour—typical of refitted vehicles. LOWER CENTRE: Ausf H with mesh skirts. BOTTOM: Original project drawing for Panther turret on PzKpfw IV (W. J. Spielberger).

8: Medium and Heavy Tank Developments, 1938-42

A NUMBER of prototypes and projects were produced to fit ever changing tactical requirements, and most of these were ultimately either cancelled or absorbed into new programmes. A summary of various miscellaneous developments which were concurrent with production vehicle output is given here.

VK 3000 series

Experience with the DW 1 and 2 designs led to a modification of requirements in 1940. A 30 ton class vehicle was postulated, with 7.5 cm KwK L/24 and provisions for a 10.5 cm KwK L/28 later. Henschel of Kassel and Porsche of Stuttgart MAN and Daimler-Benz were all asked to produce designs.

VK 3001 (H): This vehicle resembled the PzKpfw IV in its superstructure and turret, but the running gear differed radically, comprising seven medium-sized interleaved road wheels as in the DIV 2, with three return rollers each side. The vehicle had a three-speed L320 steering mechanism, a Maybach-Olvar 40 12 16 gearbox (8 forward speeds plus reverse), and an engine. Four prototypes were completed, two in April 1941 and two in October 1941. By the time the VK 3001 prototypes were ready, the PzKpfw IV had proved itself in service and the VK 3001 was not sufficient an advance to justify its adoption, since the PzKpfw IV was considered very satisfactory. In July 1941 the Soviet T-34 was encountered which had a profound effect on German tank development. Efforts were then concentrated on designs which were superior to the T-34 and the VK 3001 was too lightly armed and armoured. In the case of the VK 3001 (H), however, the first two prototypes were converted, by August 1941, into heavy self-propelled carriages with 12.8 cm guns. The vehicle was lengthened at the rear and the superstructure was open-topped. These two vehicles saw service on the Russian Front in 1942.

VK 3001(P)/Porsche Typ 100, Leopard: The Porsche design was the first AFV to be produced by Dr Porsche's design bureau. Unlike the Henschel design, the Porsche vehicle used Dr Porsche's idea of a petrol-electric drive. Two petrol motors (V-10 air cooled) mounted in parallel were linked to electric generators which in turn powered electric motors driving the rear sprockets. All speed control and steering was electric with a Voith 'Nita' torque converter. The bogie wheels were mounted in pairs with longitudinal torsion bars, one to each bogie. The engines were specially designed by Porsche and built by Graz of Vienna. Mechanically the vehicle was completely unconventional, the petrol-electric drive being Porsche's idea for powering a heavy armoured vehicle (he had used similar principals on some gun tractors he had designed for Austria in World War 1). The first two prototype vehicles were built by Nibelungenwerke:these were built and tested in 1940-41, though turrets were mock-ups, the designed armament and turrets never

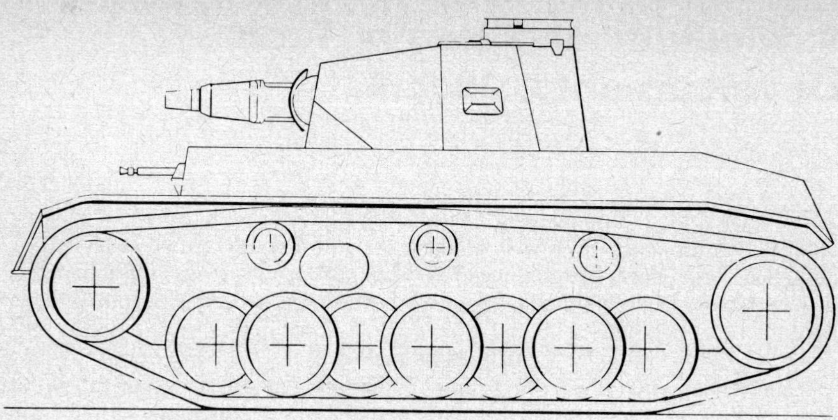

VK. 3001(H).

having been built. Trials showed the new drive to be successful except for engine trouble, where oil leaks were frequent. With the appearance of the T-34 the VK 3001(P) was abandoned, but several of the design features, including the mode of drive, were perpetrated in future Porsche designs. The name Leopard was an unofficial one provided by Porsche.

VK 3001 (DB): This design was also abandoned, but a distinctive feature was an air-cooled diesel motor MB 507, V-12 of 650 HP, which was specially developed by Daimler-Benz for use in AFVs.

VK 3001 (MAN): The MAN design appears not to have been completed in prototype form.

DW and SW Series

The DW tanks (Durchbruchswagen = breakthrough vehicle) had their origin in 1937 as a contemplated successor to the PzKpfw IV/BW series which were then in prototype stage. Though there were two designs produced, neither saw completion, the requirements changing during the development period.

DW 1: Henschel were asked to develop this vehicle at the start of 1937, the vehicle being an improved version of the PzKpfw IV in all essentials, with medium-sized interleaved wheels replacing the leaf-spring bogies used in the PzKpfw IV. One chassis was completed by 1938 and trials were carried out. The interleaved wheels were found to give a superior ride and prolonged tyre life compared the PzKpfw IV. Work was suspended in 1938 in favour of the DW 2.

DW 2: This was a heavier and improved version of the DW 1 and was seen as a replacement design for the Nb Fz/KzKpfw V/PzKpfw VI then in production. The requirement was formulated in April 1937 and the development contract was placed in September 1938. The vehicle had 50 mm of fron-

Front and rear views of VK 3001(P) Leopard on test.

tal armour, weighed 32 tons, had a crew of five and was to have the same armament as the PzKpfw IV. The turret was designed and built by Krupp. A series of eight pre-production vehicles was ordered and these were delivered early in 1940. The engine was a Maybach HL 116 6 cylinder unit of 300 HP. Top speed was 25 kph. By the time the prototypes appeared, requirements had changed and the DW designs were succeeded by the VK 3001(H).

VK 6501/PzKpfw VII /Sturmwagen (SW): This was a short-lived project for a big 65 ton assault tank (Sturmwagen) which Henshcel were asked to develop in 1938. It was provisionally designated PzKpfw VII and was an enlarged version, in essence, of the PzKpfw IV. It was to have interleaved road wheels, 7.5 cm main armament, and an auxiliary machine gun turret in the front superstructure. Two prototypes were begun and are thought to have been completed in running hull form by early 1940, but the entire project was cancelled in favour of lighter vehicles.

PzKpfw III/IV

It was appreciated from early on that there was considerable overlap of function and a duplication of effort in producing the PzKpfw III and the PzKpfw IV simultaneously. In an effort to make better use of resources, an investigation was made by Heereswaffenamt in the latter part of 1941 and the conclusion was that a merged design to be known as the PzKpfw III/IV would lead to complete standardisation of components, give better production flexibility for special variants, give better ammunition stowage compared with the PzKpfw III, allow for a larger turret and hence subsequent upgunning, and incorporate hydraulic turret traverse. Interleaved suspension was proposed. The plan was to build two vehicles with similar characteristics (50 mm frontal armour) but differing in main armament, though with common components. These vehicles would be the PzKpfw III nA and PzKpfw IV nA (nA = neuer Art = new model). Weight limit was to be 23.5 tons. It is thought that the PzKpfw III built experimentally with FAMO suspension (ZW 40) was a development model for the PzKpfw III nA within this PzKpfw III/IV programme. The ideal of a merged design was not finally realised, however, due to the ever pressing need to maintain production of existing tanks, specially the PzKpfw IV. In late 1944 all plans for this project were finally dropped. Proposed armament for the PzKpfw III nA was the 5 cm L/42 or the planned Waffe 0725 gun.

Unified PzKpfw III/IV Chassis

While the PzKpfw III/IV design was not built , there was an attempt to unify the designs as far as possible to arrive at a standard medium chassis for use in all types of AFV. In July 1944, when the economic situation was crumbling fast and resources were greatly overstretched, the Vorlaufige Richtwertprogramm IV was introduced (optimum value programme). It was proposed to combine features of the PzKpfw III chassis with those of the

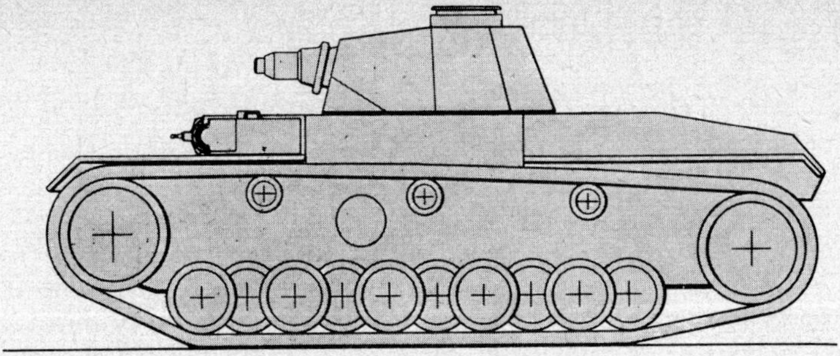

Project drawing for VK 6501.

PzKpfw IV. The resulting chassis was used exclusively for self-propelled guns and assault guns though a Sturmpanzer III/IV was provisionally planned to go into production on this chassis in early 1945. The last production model of the PzKpfw IV (Ausf J) did, however, begin to utilise components from the unified PzKpfw III/IV production programme in 1945.

Munitions-Schlepper I(A) (Sd Kfz 111).

9: Tiger Series (VK 4501)

THE Tiger tank is one of the best known of all the AFVs produced in World War 2. The development of this vehicle can be traced back to a meeting held on May 26, 1941, when Hitler suggested the need for a heavier tank than the PzKpfw IV to become the principal weapon of the panzer divisions. He had been impressed by the heavy armour of the British Matilda and French Char B tanks. Various work had already been undertaken to develop heavier tanks than the PzKpfw III and IV (eg. the DW and VK 3001) since 1937, but none of these was adopted due to the satisfaction felt with the PzKpfw III and IV designs.

Within a month, on June 22, 1941, Operation Barbarossa, the invasion of Russia had commenced. Very soon after this the first Soviet T-34 and KV-1 tanks were encountered, designs which in firepower and armour quite outclassed the PzKpfw III and IV. This experience gave a great impetus to German tank development, for there was now an urgent need to restore AFV superiority. The initial requirement was for a tank with a gun capable of penetrating 100 mm of armour at 1,500 metres, and able to withstand a hit from a comparable gun. One existing gun which could be adapted was the successful 8.8 cm Flak 36. However, Heereswaffenamt were in favour of using a smaller calibre weapon if possible, to save on size and weight of the vehicle.

There was a good case for either line of development, and one requirement was given to Henschel to produce a 36-40 ton vehicle with a tapered bore gun, Waffen 0725, while another requirement was passed to Porsche to develop a 45 ton vehicle to take a tank gun adaptation of the 8.8 cm gun. The Henschel vehicle was the VK 3601, and the Porsche vehicle the VK 4501. Krupp were to build turrets for each vehicle. The original VK 3601 project was relatively short-lived. Taper bore guns required extensive use of tungsten steel, which was in short supply, so the planned Waffen 0725 was cancelled on Hitler's direct orders. By this time VK 3601 prototypes had been built, the seventh and last being completed in April 1942. In an attempt to produce an effective tank despite the cancellation of its planned armament, the Waffen 0725, it was decided to use the Krupp turret developed for the VK 4501 on the VK 3601. This turret had a ring diameter of 185 cm as against a diameterof 165 cm on the VK 3601. In order to accommodate the bigger turret, therefore, Henschel widened out the hull superstructure to overhang the top run of the track, giving the hull a cross-section like a shallow 'T'. These changes, the heavier gun and turret, widened superstructure, and the need to widen the tracks and fit an extra run of road wheels to reduce ground pressure, all led to a big increase in overall weight, and the VK 3601 project was accordingly redesignated VK 4501(H). VK 3601 had already been built to maximum permitted width to fit a rail flat, and the fitting of the wider (72.5 cm) track presented an extra problem. One early idea was to have twin tracks each side, the outer of which could be removed for transportation, but

RIGHT: VK 3601(H) prototype (Col R. J. Icks). BELOW: Two views of an early production PzKpfw VI Tiger Ausf E with narrow 'transport' tracks.

the problem was subsequently solved by having two separate tracks, a 'battle' track of 72.5 cm width and a 'transportation' track 52 cm wide which was used with the outer run of road wheels removed.

VK 4501(P), PzKpfw VI 'Tiger'(P): Meanwhile the Porsche-designed VK 4501 was re-designated VK 4501(P) to distinguish it from the Henschel vehicle. As with the Henschel vehicle, it was stipulated when the order was given (in May 1941) that the prototype was to be ready for a demonstration before Hitler on his next birthday, April 20, 1942. Porsche utilised most of the design features from the earlier VK 3001(P) Leopard which had been cancelled. Known to Porsche as Typ 101, the VK 45001(P) featured petrol-electric drive. The engines were air-cooled and proved a constant source of trouble, so much so that Porsche contemplated switching to hydraulic transmission as an alternative to electric transmission. VK 4501(P) was demonstrated to Hitler as stipulated in a series of competitive trials with VK 4501(H), and the Henschel design was chosen as superior and was accepted for production—though study of the reports of the trials shows that there was actually little difference in performance. Undoubtedly, however, Heereswaffenamt would have been more in favour of the 'conventional' motive power system of VK 4501(H) rather than Porsche's unconventional petrol-electric drive system. Porsche, however, was a favourite of Hitler's and an order for 90 Porsche Tigers had been placed even before the trials took place. Aside from about three trials vehicles, however, none were completed as gun tanks and the 90 chassis were completed as Panzerjager 'Ferdinand' (later Elefant). Three vehicles, in turn, were completed as recovery vehicles without guns, but Elefant development is outside the scope of this book. It is a possibility (though unconfirmed) that the recovery vehicles were actually converted from the three original trials tanks.

PzKpfw VI, Tiger Ausf E(Sd Kfz 181): The VK 4501(H) was clearly more suitable for mass-production than VK 4501(P), quite apart from the virtue of its conventional engineering layout, Henschel had already made plans to produce 60 vehicles, even before the Rastenburg trials, and subsequently they prepared to build 1,300. Hitler ordered the vehicle into immediate production, and this commenced at Henschel's Mittelfeld (Kassel) factory in August 1942 at the rate of 12 a month. At Hitler's insistence this was increased, with an output of 25 a month by November 1942. Production ceased in August 1944 after 1,355 had been built, and peak output was 104 vehicles in April 1944. The name Tiger was adopted for the vehicle (this had originated as a name for the Porsche design), and the original full designation was PzKpfw VI Tiger Ausf E. From February 27, 1944, on Hitler's orders, it was known more simply as PzKpfw Tiger Ausf E, though has been unofficially called this for some time previously.

The Tiger was the most powerful tank in the world at the time of its introduction. It was of excellent quality, superbly engineered, and requiring many man-hours to produce. It weighed over 56 long tons, but was quite light to control due to a fully automatic gearbox (eight forward ratios) and fully regenerative steering. Due to the size and weight the Tiger was considered too heavy to cross most road bridges; it was therefore elaborately equipped for deep wading up to about 4 metres in depth so that it could cross the bed of

TOP: Early production Tiger Ausf E with 'battle' tracks. ABOVE: Late production Tiger Ausf E with all-steel wheels. BELOW: VK 4501(P), Tiger(P).

rivers. For submersion all hatches, vents, and other orifices could be sealed, a rubber ring was inflated round the turret ring, and a snorkel air tube was fitted to the rear decking. Fresh air passed through the fighting compartment, then into the engine compartment. The radiator compartments were flooded under water for engine cooling and clapper valves on the exhausts, together with back pressure from the engine, were relied on to keep water out at the exhaust stage. There was a bilge pump fitted to clear water seepage. This deep wading facility was rarely if ever used, and the expensive wading equipment was dropped after 495 vehicles had been built.

With well-trained crews, the Tiger was an effective tank, but when the standard of training and maintenance fell there were problems. The need to change over tracks made it unwieldy, and the range was short due to high fuel consumption of 2,3/4 gallons per mile. Slow turret traverse was another drawback; while it had hydraulic traverse equipment there was also hand traverse available to the gunner and commander. The Army wanted to build up and train a big Tiger force for the spring offensive in 1943, but Hitler was insistent that Tigers were committed to action at the first possible moment. Some were first used unsuccessfully, hampered by swamps, in a small engagement near Leningrad. Tiger battalions were used as independent units under the command of GHQ, but as the force available built up it was planned to include them in panzer division formations. However, this only happened within some Waffen-SS panzer divisions.

Due to the great weight, size, and relatively slow speed of the Tiger, it was never used as originally envisaged to spearhead the panzer divisions. Instead it was used mostly in a defensive role, in which it presented many problems to Allied forces, due to the effectiveness of its armour.

There were many production changes to the Tiger as time progresses, most of these due to rationalisation and the need to trim costs and building time. Elimination of the wading facility has already been mentioned. Air precleaners (Feifel), very necessary for North Africa and the Russian front, were fitted on all but the early models, but were discontinued from early 1944. 'S' mine dischargers were fitted in five positions on the hull top (but not on early vehicles) to give protection against infantry attacks, but these were also discontinued when production was simplified. The earliest vehicles lacked a turret escape hatch, but had two turret pistol ports. Subsequently an escape hatch at the left rear became standard. A gun loader's periscope in the turret roof was another retrospective fitting.

Major changes in production took place from early 1944 when many fittings were standardised with those of the Tiger Ausf B and Panther. The drum type cupola with vision slots was replaced by the Tiger Ausf B type cupola which had seven periscopes. The gun loader's periscope was deleted, and the 'S' mine discharges were eliminated and replaced by a Nahverteid gungswaffe smoke-bomb anti-personnel mine discharger, fitted in the turret roof, and with a 360 degree traverse. Finally, resilient steel tyred wheels replaced (from early 1944) the dished rubber tyred wheels originally fitted; at the same time the narrow transportation tracks were adopted as standard, and the wide battle tracks, plus the outer run of removable track covers used with the wide tracks.

PzBfWg Tiger Ausf E, showing extra 'star' aerial and late pattern cupola.

The first 250 vehicles had the Maybach HL 210 21.33 litre petrol engine of 642 hp, but this was replaced by the uprated Maybach HL 230 engine of 23.88 litres, giving 694 hp at 3000 rpm. Tiger Ausf B and Panther Ausf A and G also had this engine.

Tiger chassis numbers: 250001-251350.

PzBfWg Tiger Ausf E(Sd Kfz 267 or 258): This was a command version of the Tiger produced in small numbers. Externally it was similar to the standard vehicle save for extra radio aerials. Sd Kfz 267 carried Fu 5 and Fu 8 radio, while Sd Kfz 258 carried Fu 5 and Fu 7.

Bergepanzer Tiger Ausf E(Sd Kfz 185): At least one vehicle was converted to make a simple recovery vehicle, possibly more. The designation given cannot be confirmed. The adaptation involved removal of main armament, sealing of the mantlet, fixing the turret in the 6 o'clock position, adding a winch to the turret rear and a wire rope guide to the turret front. This was probably a field modification.

Ramm-Tiger(P): This was a project only, intended to utilise the chassis of the Porsche Tiger with a heavy turtle-back superstructure intended to demolish buildings in street fighting. Hitler instigated this idea, following the experience of the street fighting in Stalingrad. Porsche produced a wood model, but the design was not followed up.

Sturmtiger/Sturmpanzer VI/Sturmmoser: Some 18 Henschel vehicles were converted to this type, mounting a 38 cm RW 61 rocket projector, again intended for street fighting. They saw limited service from late 1944 but the development history is outside the scope of this book.

Projects: Further projects involving the use of the Henschel Tiger chassis included a Tiger with dozer blades for street clearing, a carrier for a 42 cm mortar, and the use of two vehicles, less turrets, to transport (in tandem) the Rhinemetall 24 cm K4 heavy field gun. None of these projects was pursued.

10: Panther Series (VK 3002)

WHILE the T-34 was first met in action in small numbers in July 1941, it was not until October of that year that the type was engaged in large numbers. 4.Pz.-Div., part of 2.Pz.-Armee, was severely mauled in an engagement at Mzensk where the T-34 demonstrated its unquestionable superiority over the PzKpfw III and IV. Development of the Tiger tank had already been spurred on in July 1941 when the existence of the T-34 and KV-1 was discovered. The October set-back, however, caused General Guderian, by now commander of 2.Pz.-Armee, to request urgent action in the shape of a Panzer Commission specially to study the T-34 problem. The Armaments Ministry sent this commission, composed of Heereswaffenamt officials and industry representatives, to the Russian front. They acted swiftly; on November 20, 1941, they were with 2.Pz.-Armee, and on November 25, they made their initial report. The main characteristics of the T-34 which rendered German tanks obsolescent were (1) sloped all round armour which gave optimum shot deflection, (2) large road wheels and wide tracks giving a stable and steady ride, and (3) an overhanging gun and turret, a feature previously avoided by the Germans. Two main armaments contractors, MAN and Daimler-Benz, were ordered to commence studies of 30 ton tank designs to incorporate these characteristics. In January 1942 the Panzer Commission reported to Hitler, and provided detailed specifications to MAN and DB.

The new design, VK 3002, was to weigh 35 tons, have a top speed of 60 kph (37 mph), a 7.5 cm gun, armour maximum of 100 mm (turret) and a minimum of 35 mm at the rear, a fording depth of 4 metres, and a Maybach HL 210 engine. The designs based on this specification were submitted only four months later, in April 1942.

VK 3002(DB): This was a most interesting prototype design which was a very close copy of the T-34 in both looks and layout. It had all steel wheels on semi elliptic leaf springs, a transverse diesel engine (MB 507), the turret mounted well forward, escape hatches in hull sides, jettisonable external fuel tanks (again like the T-34), and the driver accommodated in the turret cage with remote hydraulic steering. A wooden mock-up was made and work started on building a running prototype. This was not completed however, for the 'Panther' Committee which Wa Pruf 6 had set up to evaluate the designs decided that the more conventional engineering of the rival VK3002(MAN) design, more in accordance with previous German practice, was more suitable for production.

VK 3002(MAN): This design featured the Maybach HL 210 engine, sloped hull armour, front sprocket drive, and complex interleaved road wheels with torsion bar suspension. The turret was set well back and was based upon the Rheinmetall design produced for the VK 4501(H2) which was never proceeded with in connection with Tiger tank development. The gun was the 7.5 cm L/70 piece (originally to be L/60), also first specified for the VK 4501 (H2). The design was selected 'off the drawing board'. Top priority was given to production of a pilot model and the establishment of a produc-

TOP TO BOTTOM: VK 3002 (DB) project model. Panther Ausf D1. VK 3002 (MAN) prototype (IWM-STT7544).

tion line – in fact, the effort to get the vehicle – to be named Panther – into large scale production was the most intensive of all German wartime tank programmes. Ing Kniepkampf, chief engineer of Wa Pruf 6, took personal charge of chassis design and the interleaved road wheels and the suspension showed his influence, for he had been instrumental in developing this type of suspension for semi-tracks and some earlier tanks. A syndicate of engineering firms were made responsible for various key components, track, wheels, torsion bars, and so on. All torsion bars were to be interchangeable and it was assumed that the overall weight of the tank would not rise above 35 tons. However this proved optimistic, and a big increase in estimated weight (up to 44 tons and more) was to lead to many of the mechanical problems which beset development. Some design limitations became apparent even while testing the two VK 3002 (MAN) pilot models. Glacis plate armour was increased from 60 mm to 80 mm on Hitler's orders (though this change was not incorporated in the pilot models), but the sum effect of the gradual increase in weight above 35 tons was to put a big strain on the engine and transmission system originally planned. The Maybach HL 210 engine was replaced by a bored out version, Maybach HL 230, and a new gearbox (AK 7-200) was designed, since the original clutch and brake system could not cope. The AK 7-200 was a superior design, however, which allowed both fixed radius and neutral turns. Speed was limited to a maximum of 45 kph.

PzKpfw V, Panther Ausf D(Sd Kfz 171): Original targets to produce 250 Panthers by May 1943 were revised upwards to 600, due to the urgent need. Henschel, Daimler-Benz, and MNH, were all switched to building Panthers. The first vehicle was completed by MAN in November 1942 and the first 20 vehicles (known at one time as Ausf D1) were similar to the VK 2002(MAN) prototypes with 60 mm glacis plates, a single baffle muzzle brake on the early production L/70 gun, and the commander's cupola on the far left of the turret and bulging into the turret side. The new ZF7 (AK 7-200) gearbox and, probably, HL 210 engine, were fitted, Regular Ausf D production really got under way in January 1943 with DB joining in (and MNH and Henschel in the next two months). The main production Ausf D (known at one time as Ausf D2) had the 80 mm glacis, the cupola shifted to the right to allow a smooth turret side and simplify production, and a double baffle muzzle brake on the gun. Features of this model included: drum-shaped cupola with vision slits, pistol and shell case disposal ports in the turret sides, 'lettterbox' vertical flap for the radio-operator's hull machine gun, and a vision flap for the driver, both in the glacis, and binocular gun sight with two openings on the left side of the mantlet. The armour was stepped, morticed, and welded, with the hull superstructure overhanging the track. Separate stowage boxes were fitted beneath the side superstructure towards the rear. Crew of five were driver, radio operator, gunner, and loader, in the usual positions. Two periscopes each were provided for the driver and radio operator (hull gunner). The turret was traversed by hydraulic power, but had optional hand traverse with a common wheel for either function. There was a full turret cage. The engine and its auxiliaries were located at the rear of the vehicle, and the gearbox and transmission in the nose. There was no provision for deep wading. Skirt armour was fitted to protect the top run of the track from February 1944, though it was not always carried. The Ausf D was phased out of production from July 1943, though Henschel were still building them until

TOP: Panther Ausf D. ABOVE: Panther Ausf A. LEFT: Panther Ausf A (IWM-MH4140).

September 1943, the later vehicles often featuring Ausf A fittings. The Panther was first used in action at Operation Zitadelle, the Kursk Offensive, in July 1943, but the vehicles were affected by early 'teething troubles' which included engine over-heating and mechanical breakdowns. About 600 Ausf D were built.

PzKpfw V Panther, Ausf A(Sd Kfz 171): This new model commenced production from July 1943 and featured many detail improvements. A new cupola with armoured periscopes and rounded surfaces was fitted for the commander. A ball-mount in the glacis was fitted for the hull machine gun. Wheel construction was strengthened (wheel breakage was a big problem with the Ausf D) and the nuts in the wheel assembly were increased from 16 to 24 as a result. The binocular gun sight was replaced by a sighting telescope (hence a single aperture instead of two in the mantlet), and the pistol ports and shell-ejection port were deleted from the turret to simplify production and improve the strength of the turret walls. Problems with the engine and transmission were gradually overcome as production continued. About 1,400 Ausf A were built, becoming a major service type.

PsKpfw Panther Ausf G(Sd Kfz 171): In February 1944, the vehicle was redesignated PzKpfw Panther. This co-incided approximately with the introduction of a further refined model, Ausf G. This model resulted from work done since February 1943 on the development of new versions of the Panther (Panther II) and Tiger (Tiger II), and it was found that several of the improvements which were earmarked for the Panther II (or Ausf F) could be incorporated in the existing Panther Ausf A. The superstructure sides were altered to one-piece sheets, which eliminated the separate lower fuel tanks,simplified production, and improved protection. The thickness was increased from 40 mm to 50 mm and slope from 30° to 40°. The driver's vision hatch was eliminated and replaced by a rotating periscope. New hinged hatches with spring-assisted opening replaced the original pivoted type hatches supplied for the driver and radio operator. Late production models had new untapered mantlet fitted to overcome another problem, the downward deflection of shells which hit it, and internally there were new armoured ammunition bins with sliding armoured doors. Ammunition stowage was increased from 79 rounds to 82. There was a better method of affixing the skirt armour, and some re-stowage of equipment.

Panther Ausf G became numerically the most important of all German battle tanks in the last year of the war. Demag had joined the Panther production group and from the start of 1944 until the end of the war in May 1945, some 3,740 Panthers were built. In late 1944 it had been decided to concetrate production of tanks with traversing turrets on to only two vehicles, the Panther and Tiger II. By the end of 1944, work on the new Panther II (Ausf F) was far advanced, though it never actually entered production. However, features intended for the Panther II were incorporated into very late Panther Gs, notably the all-steel resilient road wheels (as in the late Tiger E) which replaced the dished rubber-tyred wheels previously used. Infra-red night sights and stabilisers for the main armament were also planned for the Panther Ausf G, though very few (probably only test vehicles) had them by the war's end. Also planned, but never incorporated into production vehicles, was a higher compression ratio for the engine, petrol injection, and improved cooling, bringing power output up to 800 bhp. Use of a supercharger was also

Panther Ausf A (IWM-MH4141). ABOVE: Panther Ausf G (in British service). LEFT: Turret detail, Panther Ausf D.

contemplated later, bringing output up to 900 bhp, though this modification never took place in production vehicles. Chassis numbers: 151001–250 000.

Panther Ausf F/ Panther II(Sd Kfz 141/1): In February 1943, Wa Pruf 6 asked MAN and Henschel to prepare improved designs for the Panther and Tiger respectively, to incorporate all lessons of combat and production problems previously encountered and to ensure maximum interchangeability of parts. The improved Panther, designated Ausf F, but also know as Panther II, was to have a hull similar in shape to previous Panthers, but with the same all-steel resilient wheels as the Tiger II. Many mechanical parts would be the same as those of the Tiger II. Top armour was to be doubled to 25 mm, and the hull machine gun changed from a MG 42. The major change, however, was a completely new turret design, Panzerturm Schmal (small armoured turret) with a 30% improvement in armour production and requiring 30% less time to manufacture, all within the existing weight and ring diameter limit. Daimler-Benz developed the turret as a separate project; it could take the same L/70 7.5 cm gun, a proposed lengthened L/100 7.5 cm gun, or the same 8.8 cm gun as the Tiger II. The old wide mantlet was dropped in favour of a conical Saukopf (pig's head) type, offering a much reduced frontal area. The design of the L/70 gun was modified to enable it to fit the turret, and it was possble to dispense with the muzzle brake even though the recoil forces were increased in the new turret. A built in stereoscopic rangefinder and a gyrostabiliser for the gun were standard features; also, fittings were built into the Panther II itself to enable any vehicle to be used in the command role. While the Schmalturm was built and tested in late 1944, the deteriorating situation meant that the Panther Ausf F was never actually put into production, though running prototypes were produced. One prototype is preserved at Fort Knox, Kentucky, but this has the regular Panther turret.

BefPzWg Panther (Sd Kfz 267 and 268): These were modified vehicles of either the A, D, or G production type fitted with extra radio equipment and the associated aerials for command work. Sd Kfz 267 had Fu 5 and Fu 8 equipment, while Sd Kfz 268 had Fu 5 and Fu 7. The ammunition stowage was reduced to 64.

BeobPzWg Panther (Sd Kfz 172): This was a proposal to use old Ausf D vehicles converted to provide an observation vehicle for SP artillery staffs and commanders. A short wood dummy barrel replaced the gun, the turret was fixed in place, a map table was added inside the turret, and extra radio gear was installed. A ball-mounted MG 34 in the turret front was the main armament. No production materialised, only a prototype being built.

Bergepanzer Panther (Sd Kfz 179): Early Panther actions, such as the Kursk offensive, showed the urgent need to produce a suitable recovery vehicle to bring back damaged vehicles (many had to be abandoned). The 18 ton semi-tracks used in this role were generally inadequate for towing Panthers and Tigers except when used in tandem. BgPz Panther (also known sometimes as the Bergepanther) was initially built by Henschel, but later Demag produced over 300 of them. A movable winch and winch motor were fitted in the open topped fighting compartment, a low superstructure of wood over mild steel framing was produced round the former turret opening, and there was a canvas cover for inclement weather. An A frame was fitted over the engine decking, supporting a towing eye and winch rope rollers. A heavy earth spade was hinged at the rear and was lowered and hoisted by the vehicle

TOP: BefPzWg Panther (Ausf D). ABOVE AND RIGHT: Bergepanzer Panther (IWM-STT8190).

winch. There was a light demountable jib crane which could be erected as required. Not all vehicles had the spade or winch.

Munitionspanzer Panther: Some old 'war weary' vehicles had their turrets removed, giving a superficial resemblance to the BgPz Panther. These were used, in the 1944-45 period as munitions carriers to supply units operating Panther tanks.

Projects: A mine-clearing tank (Raumpanzer), dozer tank, and an AA tank based on the Panther were projected by late 1944, but none were built. Most important project to be realised was the Panther equipped with infra-red night-fighting sights. These operated in conjunction with a 60 cm infra-red searchlight (Uhu:Eagle Owl) carried in a half-track and used to illuminate the target for the tank at 1,000-500 metres range. A few of these vehicles were produced in late 1944 and early 1945.

SP Vehicles: Most notable development of the Panther was the famous Jagdpanther, with 8.8 cm gun, a fast tank destroyer, well deployed and of major importance, though its development is outside the scope of this book. Projects included a Waffentrager based on the Panther chassis components, and a SP 10.5 cm vehicle based on the lengthened chassis of the proposed Panther Ausf F. A PzJg Panther II with 12.8 cm gun was also proposed for this chassis. None of these were built but most existed in design or mock-up form.

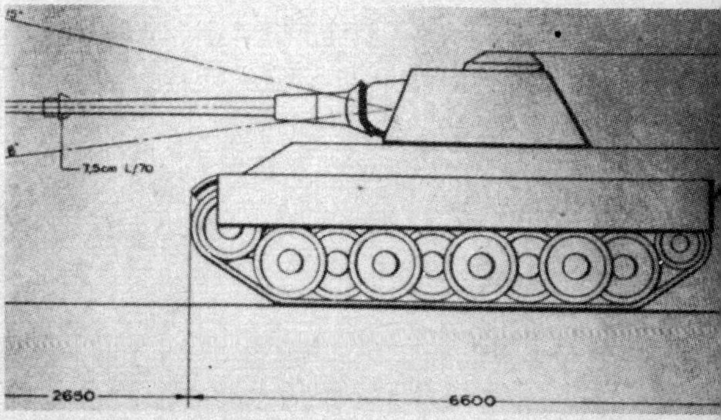

ABOVE: Project drawing for Panther Ausf F (Panther II) with alternative 7.5 cm and 8.8 cm guns. ABOVE LEFT: Schmalturm for Panther II. LEFT: BeobPzWg Panther prototype.

11: Tiger II (VK 4502)

IN August 1942, Hitler demanded that the long barelled 8.8 cm gun be fitted in the Tiger and by January 1943 he demanded 150 mm frontal armour. Heereswaffenamt asked Henschel if they could revise the design to give it fully sloped armour. By February 1943 this request was formalised into an official requirement and both Henschel — the Tiger builders — and MAN — the Panther builders were asked to produce 'second generation' replacement versions of the existing vehicles with maximum interchangeability of components. This requirement was designated VK 4502, and Dr Porsche was also asked to submit a design.

VK 4502(P)/Porsche Typ 180: The Porsche design existed only in project form. This offered two alternative layouts with the turret either at...the forward or backend of the hull, The vehicle with rear-mounted turret was provisionally accepted. The hull itself was compact, with sloped faces like the Panther. As with previous Porsche designs petrol-electric drive was offered, but the design was ultimately rejected by Wa Pruf 6, partly because they preferred conventional drive but also because copper, used extensively in the electric motors, was in short supply. The suspension included paired bogies with longitudinal torsion bars similar to previous Porsche designs. The turret design was put in hand for this vehicle by Wegmann. The main gun was to be the 8.8 cm KwK 43 L/71; some 50 of these turrets were put into production before the VK 4502(P) was abandoned.

VK 4502(MAN): This project was accepted and design and production plans went ahead. The production vehicle was to be PzKpfw Panther II and is described in the previous section. It did not actually enter production before the war ended.

VK 4503(H): The Henschel design was a logical development of the Tiger Ausf E. It had sloped armour, a large turret to carry the 8.8 cm KwK 43 L/71, and was powered and driven conventionally. The design was accepted and was ordered into production as a top priority project. Henschel did try to have a less drastic development undertaken as the VK 4502(H)—but this was rejected, the revision of this design becoming VK 4503(H).

PzKpfw Tiger Ausf B/TigerII/Koenigstiger(Sd Kfz 182): The actual detail design of the vehicle based on the VK 4503(H) project was not completed until October 1943, three months behind the original schedule, due to the extensive redesign and the extra work needed to standardise components with the Panther II. The pilot model was ready in November 1943, however, and production began in December at the Henschel factory, Kassell, adjacent to the Tiger Ausf E line. The first production Tiger Ausf Bs were ready in February 1943, and the first 50 were given the Wegmann turrets which had been originally designed for the abandoned VK 4502(P). All other Koenigstiger built, however, had a new Krupp designed turret which was simpler to build and avoided a shot trap under the mantlet which was a shortcoming of the Porsche design. Tiger Ausf B, which was also known as Tiger II, Koenigstiger (King Tiger), or Royal Tiger to the Western Allies, had many of its features included on the later production Tiger Ausf E, though it more closely resembled the Panther externally. Engine, cupola, engine covers were common to the late Tiger E and later Panther, and the all-steel resilient steel wheels were common to the late production Tiger E and the very late production Panther Ausf G. The Tiger Ausf B had hull armour up to 150 mm thick and was larger and heavier than previous vehicles. There was a small conical Saukopf (pig's head) mantlet, the 8.8 cm KwK 43 L/71 gun, ball-mounted hull machine gun and overlapped (rather than interleaved) wheels. This was to overcome the problems of wheel changing, and jamming by ice and compacted mud encountered with the Tiger Ausf E and Panther.

The rationalisation programme of fall 1944 provided for production of only two tanks with traversing turrets in 1945, the Panther Ausf G (or Panther II), and the Tiger Ausf B. Tiger Ausf B was complicated to build, however, and as two Panthers could be built for every one Tiger Ausf B, production was concentrated on the Panther. Production was scheduled at 100 a month, increasing to 145, but the best ever output was 84 in August 1944, failling to only 25 in March 1945. Tiger Ausf B was the heaviest operational tank to see service in World War 2. It was first in action on the Eastern front in May 1944, and on the Western front in August 1944. It was heavy (68 tons) and unwieldy with serious tactical limitations. It had short engine life and unreliable transmission. By the time it was in service, however, German tanks were largely used defensively, and the shortcomings were less serious than they might appear. Chassis numbers: 280001 - 280404 (first 50 with Porsche turret, the rest with Henschel (Krupp) turret).

BfPzWg Tiger Ausf B: This was a commander's version differing only in radio equipment and associated antenna. Relatively few were built.

SP projects: A tank destroyer based on the Tiger Ausf B was produced designated PzJg Tiger Ausf B (Sd Kfz 186). At 76 tons it was the largest and heaviest AFV to see service in World War 2; some 70 were built out of an order for 150 in 1944-45. The development story is outside the scope of this book. One vehicle was fitted with Porsche suspension.

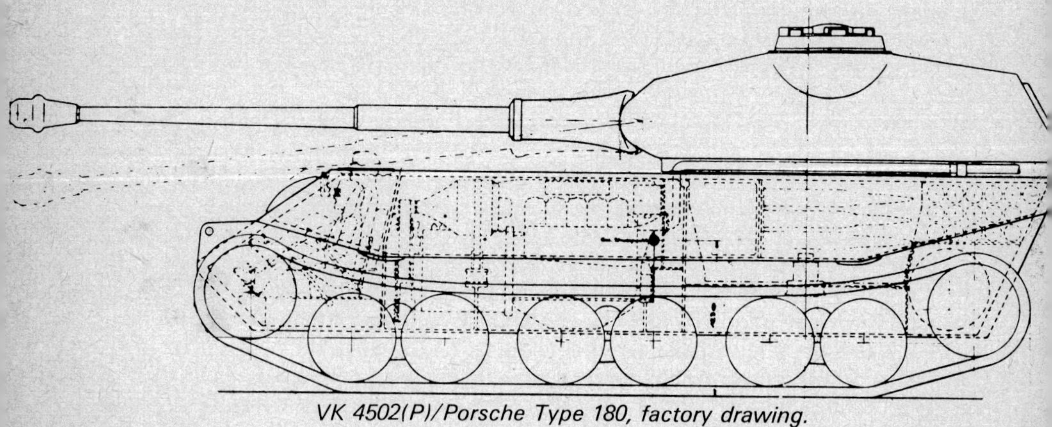

VK 4502(P)/Porsche Type 180, factory drawing.

TOP: Tiger Ausf B with standard turret (BA-721-328-27a). ABOVE AND BELOW: Tiger Ausf B with Wegmann turret.

12: Maus and Super-Heavy Tanks

DR FERDINAND Porsche enjoyed the respect of Hitler from pre-war days. Porsche had designed the Volkswagen car in the 1930s but had been involved in military vehicle design in World War 1. His design bureau offered several tank projects during the World War 2 period to meet the major Heerenwaffenamt requirements, but none of these designs was accepted, though vehicles already laid down to the VK 4501(P) design were converted to the PzJg Ferdinand(Elefant) and saw service in this form. Porsche was a great exponent of petrol-electric drive for tracked fighting vehicles — that is say, petrol engines drove generators which in turn powered traction motors which drove the tracks. While this was not a Porsche innovation (it had been tried in French and American tanks in World War 1), it was sufficiently novel to cause this mode of traction to be viewed with suspicion by the engineers of Wa Pruf 6 who much preferred 'conventional' mechanical drive. Additional factors were the lack of development time — all World War 2 projects were urgent ones — shortage of materials (copper was used extensively in electric transmission), and lack of large scale production experience for heavy petrol-electric drive vehicles.

However, Porsche succeeded in interesting Hitler, and later the Reichs Armaments Minister Speer, in a project for a super-heavy tank which would be invulnerable to all known tank and anti-tank guns. The project, under the code name Maus (Mouse) was first mooted late in 1941 and in June 1942 he got permission from Hitler to go ahead and arrange the building of a prototype. At this time Porsche's stock was low with Waffenamt and he had been removed as head of the Panzer Kommission. Porsche shared Hitler's enthusiasm for tanks with heavy armour and big guns. In his memoirs, Speer infers that Porsche was placating Hitler with Maus, but it seems possible that Hitler for his part thought that the Maus project would recompense Porsche for the rejection of his other projects by Waffenamt.

Maus: The Maus was an extravagant design, which showed a major divergence of ideas between the engineer with no practical tactical experience and the tank men in the field. Initially known under a code name of Mammut (Mammoth), before reverting back to Maus, the project got under way with the construction of a running prototype which was completed in November 1943, the Krupp firm having the building contract for the turret and guns, and Alkett for the rest. The designed weight was over 180) tons, with turret alone weighing 50 tons, and the armament was to be a 12.8 cm Pak 44 L/55 gun as fitted to the Jagdtiger. A 7.5 cm KwK 44L/36 gun was to be mounted co-axially. Frontal armour was over 230 mm thick and a special engine of 1,200 hp was designed by Daimler-Benz. It was tested in both petrol (MB 509) and diesel (MB 507) forms in the first two of the Maus prototypes. Transmission was electric, with generators beneath the turret floor. There were 6 bogies each side, each with four staggered wheels on sprung suspension assembles. The drive was to the rear sprockets with traction motors at the rear of

ABOVE: Maus prototype on test, without armament. LEFT: Maus production hull, with 12.8 cm gun barrel (W. J. Spielberger).

the vehicle, the Daimler V-12 engine being at the front. There was an 8 hp two-stroke engine to act as starter motor and drive the various auxiliaries. The road speed of the vehicle was 20 kph. The hull took the form of a heavily armoured box which overhung the tracks. The Armaments Ministry had asked that there be a maximum vertical armour thickness of 350 mm and that the belly be impervious to mines. The turret ring diameter was 300 cms, and because of its great weight a separate electric motor was required just to traverse the turret. The turret front was cast but all other armoured parts were of rolled plate. Small cranes were to be provided inside the turret to assist in loading, due to the weight of the ammunition.

Because no bridge would support a vehicle of this size, it had to be made to cross waterways by deep submersion, up to 8 metres, and snorkel trunking was to be provided. To power the electric drive under water a generator truck paying out electric cables would attend Maus from the river bank. It was subsequently planned to fit a 15 cm KwK 44 L/38 gun in place of the 12.8 cm weapon, but none of the vehicles was completed with armament. The prototype vehicle, with a weighted simulated dummy turret, ran on December 23, 1943 and trials continued until May 1944 at Boblingen. In June 1944 this prototype was actually completed with the Krupp turret and armament. Two more vehicles were under construction in the last year of the war and components for others — possibly totalling as many as ten — were being built. Maus was of very limited tactical value, due to its immense size, and work on it appears to have been halted some months before the war ended.

VK 7001(K)/Lowe/Tiger-Maus: Heeres waffenamt gave Krupp a contract to produce a design in the 70 ton class, as an 'official' Army rival to the Maus design. The design was drawn up and produced as a wooden small scale mock-up. It was based on Tiger Ausf B (Tiger II) components, but had the turret set to the rear and engine placed in the centre. The mock-up shows provision for battle and transportation tracks as in the Tiger Ausf E.

Krupp-Maus projects: Various other design studies for super-heavy tanks were found at the Krupp works when the plant was captured in 1945. Vehicles in the 110, 130, 150, and 170 ton class were contemplated, one of them to be armed with a 30.5 cm mortar. Another study showed a 1,500 ton vehicle with a 80 cm gun and twin turrets carrying 15 cm guns. Four U-boat engines were to power this giant vehicle which would have 250 mm frontal armour.

PzKpfw IX and X: Issue 3 of the German magazine 'Signal' in 1944, carried illustrations of two super-heavy tanks commenting 'Almost like landships to look at, these new projects were under development — details are still to be released', and much more in a similar optimistic style. 'Signal' existed for propaganda purposes, and it seems likely that the 'PzKpfw IX' and 'PzKpfw X' were simply an artistic and journalistic impression of the Porsche-Maus and E.100 projects then under development. They are included here for completeness.

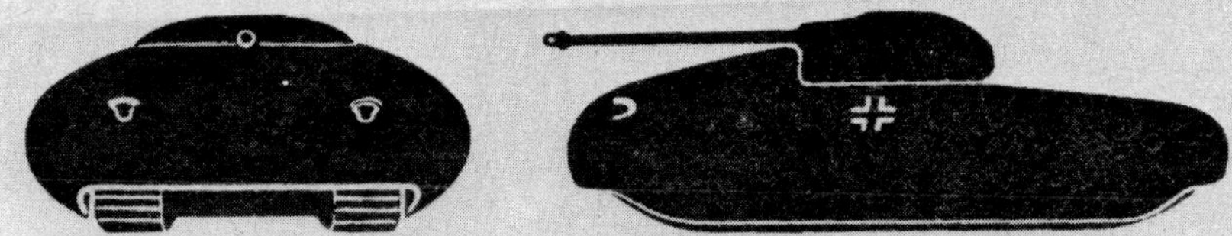

Pz. Kpfw. IX

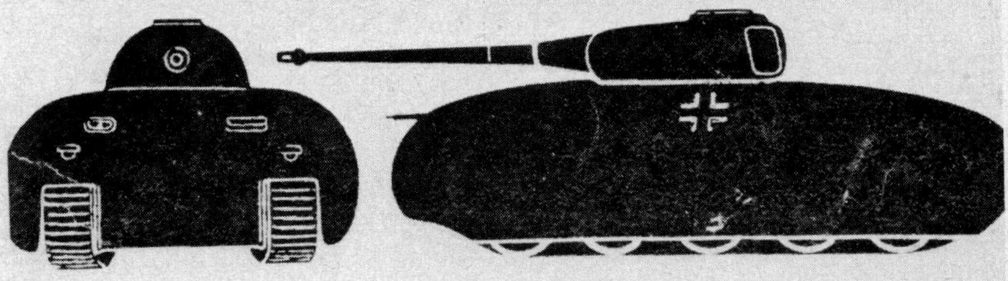

Pz. Kpfw. X

TOP: VK 7001 (K). Tiger-Maus project model. ABOVE: Impressions of 'PzKpfw IX' and 'PzKpfw X'. RIGHT: 'Signal' impression of 'PzKpfw IX'.

13: E-Series Vehicles

AFTER Porsche left the Panzer Kommission, another prominent engineer, Heydekampf, took his place as chairman. Under his direction a long term plan was drawn up to produce a range of tanks and other tracked vehicles which would be fully rationalised and make use of standardized components. The idea was to speed up production, maintenance, and provision of spares, and to cut down on costs. This was known as the Entwicklungstypen (development type) programme, or E series. Types planned were as follows:

E.10 10-15 tons light tank.
E.25 25-30 tons tank.
E.50 50 tons medium tank as Panther replacement.
E.75 75-80 tons tank as a Tiger replacement.
E.100 130-140 tons super-heavy type.

All these, except E.100, were to have rear sprocket drive, single guide tracks, and single tooth sprockets for simplicity; previous German tanks mostly had double guide tracks and sprockets. Belleville washer-type suspension developed by Adlerwerke was to replace the torsion bar suspension previously used, the advantage of this being that it was affixed entirely externally.

However, of the entire series, work largely resulted in progress to produce a prototype of the largest vehicle, E.100. Early discussions on super-heavy tanks had taken place in November 1941 when Porsche has first suggested Maus to Hitler. In mid 1943 when work commenced on Maus, Heereswaffenamt asked Adler to start on the E.100 and work on this proceeded slowly until the end of the war in May 1945, even though Hitler ordered that work on super-heavy tanks was to be suspended in late 1944. The entire project was held up, however, by the slow supply of components and vital parts. At one time only three men were available to carry out assembly work. When the E.100 prototype was captured incomplete by the Allies in May 1945, it was yet to be placed on its track, and some suspension components were still awaited. The gun was to be a 15 cm weapon and the turret, by Krupp, was exactly the same as that being produced for Maus. It was planned to fit a standard Tiger engine (Maybach HL 230) for tests, but a developed version (HL 234) was scheduled for production vehicles. Olvar transmission, and other Henschel designed components from the Tiger programme were also used. The vehicle was to have two sets of tracks—battle tracks 1000 mm wide and transportation tracks 550 mm wide. The wheels were to be of the all-steel resilient type similar to those produced for the Tiger Ausf B. The outer run of road wheels and the outer section of the front drive sprockets were removed when transportation tracks were fitted. Very heavy armoured curved side shields were to be attached to the vehicle sides when battle tracks were fitted. A jib and winch could be bracketed to the turret sides to allow the shields to be lifted off, but a crane vehicle would probably have been used

Uncompleted E.100 prototype on Gotha 68 ton tank transporter trailer.

when possible for this heavy task. When stripped for transportation the vehicle only just cleared the German railway loading gauge limits.

Frontal armour was 200 mm at 45 degrees. The basic hull shape was similar to that of the Tiger Ausf B, as was the mechanical layout. Even with the uprated Maybach engine, however, and a newly developed transmission/steering combination (Mekydro), the power/weight ratio of 8.5 hp/ton was inadequate for a vehicle this size. The turret, scheduled for either the 15 cm gun or a 17.4 cm gun, would have had a limited traverse in all probability since the turret ring was limited to 300.6 cm by the width of the hull.

While E.100 was technically and, in practical terms, a far superior answer to the super-heavy tank problem than Maus, it still represented a waste of time and resources due to the very limited tactical relevance of such a vehicle. The scale drawing shows it as would have looked completed, while the photograph shows the stage of completion it had reached by May 1945.

Of the other E series projects, a design for the E.10 is thought to have been drawn up by KHD, for the/E.25 by Argus, and the E.50 by Argus and/or Auto-Union; in each case these firms worked closely with Wa Pruf 6.

BELOW: Authors' impression of E.50 based on photographs of an unidentified running prototype of 1944-45 thought to be this vehicle. The E.50 would probably have carried the Schmalturm as designed for the Panther II. Scale drawing of E.100 is on page 160.

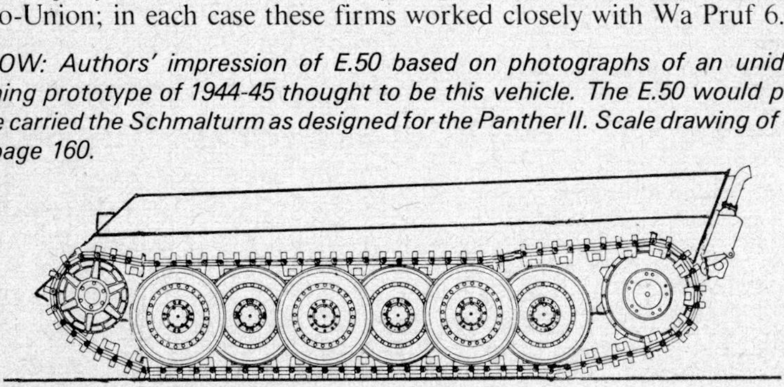

14: Specifications

IN this section specifications are given for representative models from each of the standard production types. Space limitations preclude the inclusion of date covering every single production model. All measurements given are in metric units. Metric to Imperial conversion: 2.5 cm (25 mm) to 1 inch; 1 Kilogram to 2 lb 3¼ oz; 1 metric ton to 2204 lbs; 1 litre to 1 pints. The German standard angles in which armour inclination is given took 0 degrees as horizontal, 90 degrees as vertical.

Neubaufahrzeuge

Crew: 7
Weight: 23,000 kg.

Performance

Max Speed: 30 Kmph

Dimensions

Length overall: 6.6 metres
Width: 2.9 metres.
Height: 2.98 metres.
Track width: 380 mm.

Armour

Front: 20 mm.
Side: 13 mm.
Rear: 13mm.

Mechanical Details

Engine; 1 x Maybach HL 108 TR, 12 cylinder, 280 bhp.
Transmission: 5 forward gears, 1 reverse.

Armament

1 x 7.5 cm KwK L/23.5, 1 x 3.7 cm KwK L/45 (co-axial) 4 x MG 13 7.92 mm.

Panzerkampfwagen I Ausf A and D (Sd Kfz 101)

(second figure throughout refers to Ausf B)
Crew: 2.
Weight: Combat Loaded 5.4 tons (6 tons).

Performance

Speed, Max. Road: 37 Kmph (40 Kmph).
Max. Gradient: 30 degrees.
Fording depth: 58 cm.
Trench crossing: 140 cm.
Step: 36 cm.
Min. turning circle: 2.1 metres.
Range, Road: 145 km (140 km).
Cross-country: 97 km (115 km).
Powert to weight ratio: 19.5 HP/ton (16.7 HP/ton)

Dimensions

Length overall: 402 cm (442 cm)
Width: 206 cm.
Height: 172 cm.
Ground clearance: 29 cm.

Trackwork

Length on ground: 247 cm.
Width: 28 cm.
Links per track: 84 (100).

Mechanical Details

Engine: (Ausf A) Krupp M305, air cooled 3.46 litres, 57 hp at 2500 rpm.
(Ausf B) Maybach NL 38TR, water-cooled 3.79 litres, 100 hp at 3000 rpm.
Fuel: 145 litres.
Steering: Clutch and Brake.
Suspension: Modified Horstmann.
Gearbox: ZF Aphon FG 35, 5 forward, 1 reverse.
Transmission: Sliding pinion type.

Armament

Main: 2 x 7.92 MG 13.
Traverse: 360 degrees, hand operation.

Stowage

Ammunition: 1525 rounds.

Armour

Hull, nose: 13 mm at 63 degrees.
front: 13 mm at 68 degrees.
side: 13 mm at 90 degrees (bottom).
and 73-82 degrees (upper).
rear: 13 mm at 50-75 degrees.
roof: 6mm at 0-58 degrees.
floor: 6 mm at 0 degrees.

Turret, front: 13 mm at 80 degrees.
sides: 13 mm at 68 degrees.
roof: 8 mm at 0-18 degrees.

Panzerkampfwagen II Ausf F (Sd Kfz 121)

Crew: 3
Weight: Combat loaded: 9.5 tons.

Performance

Speed, Max. Road: 40 Kmph.
Max. Gradient: 30 degrees.
Fording depth: 92.5 cm.
Trench crossing: 170 cm.
Step: 42 cm.
Min. Turning circle: 4.8 metres.
Ground pressure: 0.76 Kg/sq.cm.

Range (internal fuel) Road: 125 Km.
Cross country: 125 Km.
Power to weight ratio: 14.75 Metric HP/ton.

Dimensions
Length: 481 cm.
Width: 228 cm.
Height: 202 cm.
Ground clearance: 34.5 cm.
Fire height of gun: 159.5 cm.
Turret ring dia: 116.8 cm.

Trackwork
Centres: 188 cm.
Length on ground: 240 cm.
Width: 30 cm.
Pitch: 9.21 cm.
Number of links per track: 106/107.

Mechanical Details.
Engine: 1 x Maybach HL 62 TR, 6-cylinder inline, water cooled petrol engine of 6.191 litres developing 140 bhp at 2600 rpm.
Transmission: ZF SSG 46 gearbox with 6 forward and 1 reverse speeds.
Steering: Epicyclic clutch and brake.
Suspension: Five equally spaced rubber tyred bogie wheels on each side with quarter elliptic leaf springs.

Armament
Main: 1 x 2 cm KwK 30 or 38.
Calibre, and length in calibres: 20 mm (0.78 ins), L/97.16.
Traverse: Turret 360 degrees. Operation: Hand.
Elevation: plus 20 degrees, minus 9.5 degrees.
Secondary armament: 1x 7.92 mm MG 34.

Stowage
Ammunition, main armament: 180 rds.
secondary armament: 255 rds.
Internal Fuel capacity: 170 Litres.

Armour
Type:
Hull, Nose upper: 35 mm at 77 degrees.
Nose lower: 20mm at 16 degrees.
Glacis plate: 20 mm at 90 degrees.
Drivers plate: 30 mm at 80 degrees.

Sides: 20 mm at 90 degrees.
Rear: 15 mm at 81 degrees.
Deckling: 10 mm at 0 and 8 degrees.
Belly: 5 mm at 0 degrees.
Turret, Front: 7.176" (30 mm) rounded.
Sides and rear: 0.588" (15 mm) at 68 degrees.
Roof: 0.384" (10 mm) at 0 + 16 degrees.

Vision
Sights: TZ F4

Communications
1 x FU 5, 10 watt transmitter ultra short-wave receiver 27200-33300 Kc/s
Range: WT-2.5 miles (4 Km) RT-1.25 miles (2 Km)

Panzerkampfwagen 38(t) and 35(t)
(second figure throughout refers to PzKpfw 35(t))
Crew: 4 (4).
Weight, Combat Loaded: 9.7 tons (10.5 tons).

Performance
Speed, Max Road: 56 Kmph. (40 Kmph).
Max. Gradient: 30 degrees (28.6 degrees).
Fording depth: 91.5 cm (79.3 cm).
Trench crossing: 186 cm (186 cm).
Step: 79.3 cm (79.3 cm).
Min. turning circle: 4.54 metres (4.88 metres).
Range, Road: 201.25 km (193.2 km).
Cross-country: 143.3 km (116 km).

Dimensions
Length overall: 454.45 cm (453.4 cm).
Width: 205.43 cm (214.27 cm).
Height: 236.5 cm (219.76 cm).
Ground clearance: 40.23 cm (35 cm).

Trackwork
Length on ground: 291.38 cm (326 cm).
Width: 28.75 (26.25 cm).
Links per track: 87-93 (105)

Mechanical Details
Engine: (PzKpfw 38) 6 cylinder in-line water cooled Praga, 125-150 hp at 2,200-2,600 rpm.
(PzKpfw 35) 6 cylinder in-line OHV Skoda T-11, 120 hp at 1,800 r.p.m.
Fuel: 172.5 litres (119.68 litres)
Steering: Epicyclic clutch and brake (servo-pneumatic clutch and brake).
Suspension: Semi-elliptic spring (paired bogies on semi-elliptic springs).
Gearbox: Praga-Wilson TN-100, 5 forward, 1 reverse (Praga-Wilson-Skoda 6 forward, 6 reverse).
Transmission: Epicylic (servo-pneumatic)

Armament
Main: Skoda A7, 37.2 mm L/47.8 (Skoda A3, 37.2 mm L/40).
Traverse: 360 degrees, hand-wheel or free.
Elevation: 6 degrees, + 12 degrees (–10 degrees, 25 degrees).
Secondary: 2 x 7.92 mm 71665 CZ 37.

Stowage
Main ammunition: 90 rounds (HE, AP).
MG ammunition: 2,550 rounds.

Armour
Hull, nose: 25 + 25 mm (35 mm).
front: 25 + 25 mm (28 mm).
glacis: 25 mm (28 mm).
sides: 15-30 mm (16-24 mm).
roof: 10 mm (20 mm).
floor: 8 mm (12 mm).
rear: 12 mm (20 mm).
Turret, mantlet: 25 mm (35 mm).
front: 25 + 25 mm (28 mm).
sides: 15-30 mm (16 mm).
rear: 15-22 mm (16 mm).
10-15 mm (16 mm).

Panzerkampfwagen III Ausf M. (10/ZW) (Sd Kfz 141/1)
Crew: 5.
Weight, Combat loaded: 21.13 tons.

Performance
Speed, Max. Road: 40 Kmph.
Cross Country: 19 Kmph.
Max. Gradient: 30 degrees.
Fording depth: 130 cm.
Trench crossing: 220 cm.
Step: 60 cm.
Min. Turning circle: 5.85 metres.
Ground pressure: 1.03 Kg/sq. cm.
Range (internal fuel), Road: 155 Km.
Cross country: 95 Km.
Power to weight ratio: 11.5 Metric HP/ton.

Dimensions
Length overall: 655 cm.
Length: 578 cm.
Width with apron: 341 cm.
Width: 297 cm.
Height: 250 cm.
Ground clearance: 38.5 cm.
Fire height of gun: 190 cm.
Turret ring dia: 150 cm.
Road wheel dia (overall): 53.3 cm.

Trackwork
Centres: 251 cm.
Length on ground: 286 cm.
Width: 40 cm.
Number of links per track: 93.

Mechanical Details
Engine: 1 x Maybach HL 120 TRM Ausf A. V.12 (60 degrees) OHV watercooled Petrol of 11.867 litres developing 265 bhp at 2600 rpm and 300 bhp at 3000 rpm.

Transmission:
ZF SSG77WrA 705 with 6 forward and 1 reverse ratio.
Steering: Clutch and Brake.
Suspension: One torsion bar to each road wheel, swing arms act rearwards on left hand side, forward or right hand slide.

Armament
Main: 1 x 5 cm KWK 39
Calibre, and length in calibres: 50 mm (1.968 ins), L/60
Traverse: 360 degree turret. Operation: Hand.
Elevation: plus 20 degrees, minus 10 degrees to front and sides, minus 6 degrees to rear.
Secondary armament: 1 x 7.92 MG 34 coaxial
1 x 7.92 MG 34 Hull.
6 x 9 cm Nb K.wg-smoke dischargers.
1 x 27 mm signal pistol.

Sights
Main: TZF 5 e. Secondary: TZF 2.

Communications
1 x FU 5.-10 watt transmitter c Ultra shortwave receiver e. operating the 27200-33300Kc/s band.

Stowage
Ammunition, main armament: 5 cm 98 rds.
Ammunition, secondary armament: 7.92 mm-17 x 150 rd belts.
Signal: 24 rds.
Internal Fuel capacity: 70.44 Imp. gals. 84.54 U.S. gals. 320 litres.

Armour
Hull, Nose upper: 50 mm at 40 degrees.
Nose lower: 50 mm at 70 degrees.
Glacis plate: 25 mm at 6 degrees.
Drivers plate: 50 + 20 mm at 81 degrees.
Sides, 30 mm at 90 degrees.
Apron: 10 mm.
Rear, upper + lower 50 mm at 73 + 81 degrees.
Decking: 18 mm at 0 degrees.
Engine covers: 18 mm at 2 + 11 degrees.
Belly, Front: 16 mm at 0 degrees.

Turret Front: 57 mm at 75 degrees.
Sides: 30 mm at 65 degrees.
Rear: 30 mm at 78 degrees.
Roof: 10 mm at 7 + 0 degrees.
Mantlet: 50 + 20 mm rounded.
Cupola: 45-55 mm.

Variation for Panzerkampwagen III. Ausf L (9/ZW)
Weight: Combat loaded: 22.3 tons.
Fording depth: 80 cm.
Length overall: 641 cm.
Length: 552 cm.

Variation for Panzerkampfwagen III. Ausf N. (11/ZW)
Weight: Combat loaded: 21.3 tons.
Length overall: Production Ausf. N. 578 cm.
Converted Ausf. L. 552 cm.

Armament
Main: 1 x 7.5cm KwK.
Calibres length and calibre: L/24 (7.5cm) (2.952 ins)
Elevation: plus 20 degrees, minus 8 degrees 17 minutes.
Sights: 1 x TZF 5f.

Stowage
Ammunition, main armament: Production Ausf N-7.5cm
64 rounds. Converted Ausf. L-7.5 cm 56 rounds.
Secondary: 7.92mm 23 x 150 round belts.

Panzerkampwagen IV Ausf D (Sd Kfz 161) and Ausf G (Sd Kfz 161/1).
(second figure throughout is for Ausf G)
Crew: 5.
Weight, Combat Loaded: 20 tons (23.6 tons).

Performance
Speed, Max. Road: 40 Kmph.
Cross-country: 20 Kmph (16 Kmph).
Max Gradient: 30 degrees.
Fording depth: 80 cm (100 cm).
Trench crossing: 230 cm (220 cm).
Step: 60 cm.
Min. turning circle: 5.92 metres.
Range, Road: 200 km.
Cross-country: 130 km.
Power to weight ratio: 15 HP/ton (12.7 HP/ton).

Dimensions
Length overall: 591 cm (663 cm).
Length (excluding gun): 591 cm (591 cm).
Width: 285 cm (288 cm).
Height: 268 cm.
Ground clearance: 40 cm.

Trackwork
Length on ground: 352 cm.
Width: 38 cm (40 cm).
Links per track: 101 (99).

Mechanical Details
Engine: Maybach HL 120 TRM V-12
water-cooled petrol, 300 hp at 300 rpm.
Fuel: 470 litres.
Steering: Clutch type.
Suspension: Bogie type.
Gearbox: Synchron ZF SSG 76, 6 forward and
1 reverse gear. (ZF SGG 77).

Armament
Main: 1 x 7.5 cm KwK 24 (1 x 7.5cm KwK 40 L/43).
Secondary: 2 x 7.92 mm MG 34.

Stowage
Main ammunition: 80 rounds (87 rounds).
MG ammunition: 2,700 rounds (2,250 rounds).

Armour
Hull, nose: 30 mm at 78 degrees (50 mm at 78 degrees)
front: 20 mm at 81 degrees (50 mm at 80 degrees)
side:20 mm at 90 degrees (30 mm at 90 degrees)
rear: 20 mm at 78-90 degrees.
roof: 11 mm at 0 degrees.
floor: 10 mm at 0 degrees.

Turret, front: 30 mm at 79 degrees (50 at 79 degrees)
side: 20 mm at 64 degrees (30 mm at 64 degrees)
rear: 20 mm at 74 degrees (30 mm at 74 degrees)
top:10 mm at 0-6 degrees.

Panzerkampfwagen 'Tiger' Ausf, E. Sd Kfz 181.
Crew: 5
Weight, Combat 5 tons.

Performance
Speed, Max. 37 Kmph
Cross country: 19 Kmph.
Max Gradient: 35 degrees.
Fording depth: 122 cm or 396 cm.
Trench crossing: 229 cm.
Step: 79 cm.
Ground pressure: (Battletracks) 1.05 Kg/sq./cm.
Range Road: 117 Km.
Cross country: 67 km.
Power to weight ratio: (Gross) 12.3 metric HP/ton.

Dimensions
Length overall: 846 cm.
Length: 632 cm.
Width overall: 373 cm.
Width for transport: 315 cm.
Height to top of Cupola: 290 cm.
Ground clearance: 43 cm.
Fire height of gun: 185 cm.
Turret ring dia: 185 cm.
Road wheel dia. (overall) 80 cm.

Trackwork
Centres: 282 cm.
Length on ground: 401 cm.
Width: Battletracks: 72.5 cm.
Transport tracks: 52 cm.
Pitch: 13.3 cm.
Number of links per track: 95
Type: dry, single line

Mechanical Details
Engine: 1 x Maybach HL 210 P45 V12. Watercooled Petrol of 21.33 litres developing 642 bhp at 3,000 rmp (First 250 vehicles) or HL 230 P45 of 23.88 litres developing 694 bhp at 3,000 rpm.
Transmissions: Maybach Olvar preselective, 8 forward 4 reverse.
Steering: Henschel L600 C Regenerative (2 radii) controlled differential, hydraulic operated.

Suspension: Torsion Bar, 8 Stations per side of 3 or 2 wheels each. Hydraulic shock absorbers on front and rear axle, housed in hull.

Armament
Main: 1 x 8.8 cm KwK 36
Calibre, and length in calibres: 88 mm L/56.
Traverse: 360 degrees. Operation: Two speed, Hydraulic and hand, turret.
Elevation: plus 17 degrees, minus 6.5 degrees.
Secondary armament: 1 x 7.92 mm MG 34 co-axial.
1 x 7.92 mm MG 34 in bow gimbal mount
(Optional on late models, 1 x 7.92 mm MG 34 on ring of cupola)
1 x 9 mm MP 38 or 40.
11 x Smoke bomb launchers (replaced on late model by one Nahverteidigungswaffe)

Stowage
Ammunition, Main Armament: 92 rds. AP and HE
Ammunition, secondary armament: 7.92 mm-5100 rds in 150 rd belts.
Internal fuel capacity: 567 litres.

Armour
Type: Rolled homogenous plate interlocking stepped joints welded construction.
Hull: Nose upper: 102 mm at 66 degrees.
Nose lower: 62 mm at 30 degrees.
Glass plate: 62 mm at 10 degrees.
Divers plate: 102 mm at 8 degrees.
Sides, Upper: 82 mm at 90 degrees.
Lower: 62 mm at 90 degrees.
Rear Plates: 82 mm at 82 degrees.
Decking: 26 mm at 0 degrees.
Belly: 26 mm at 0 degrees.

Turret: Front: 102 mm at 80 degrees.
Turret wall: 82 mm at 90 degrees.
Roof: 26 mm at 9 degrees.
Mantlet 10 mm.

Panzerkampfwagen Panther Ausf G (Sd Kfz 171/1)

Crew: 5

Weight, Combat loaded: 45.3 tons.

Performance

Speed, Max. Road: 45.7 Kmph.
Cross country: 25/30 Kmph.
Max. Gradient. 30 degrees
Fording depth: 190 cm
Trench crossing: 245 cm
Step: 90 cm
Min. Turning circle: 10 metres.
Ground pressure: 0.87 Kg/sq.cm.
Range (internal fuel), Road: 200 Km.
Cross country: 100 Km.
Power to weight ratio: Gross 15.6 Metric HP/tonne

Dimensions

Length overall: 886 cm.
Length: 693.5 cm.
Width overall: 327 cm.
Height: 299.5 cm.
Ground clearance: 56 cm.
Fire height of gun: 230 cm.
Turret ring dia: 165 cm.
Road wheel dia (overall) 86 cm.
Road wheel width: 10 cm.

Trackwork

Centres: 268 cm.
Length on ground: 392 cm.
Width: 66 cm.
Number of links per track: 87. Type: dry, single pin

Mechanical Details

Engine: 1 x Maybach HL 230 P30. V.12 (60 degrees) OHV.
Water cooled Petrol of 23.88 litres developing 700 bhp (metric) at 3000 r.p.m.6.8 to 1 compression ratio 23 volt Electrical system.
Transmission: ZF AK 7-200
Steering: Man Multi-geared
Suspension: 2 transverse torsion bars per roadwheel.

Armament

Main: 1 x 7.5 cm Kwk 42
Calibre, and length in calibres: 75 mm L/70
Traverse: 360 degree turret Operation: Power take-off and hand.
Elevation: plus 20 degrees, minus 4 degrees.
Secondary Armament: 1 x 7.92 mm MG 34.
(Co-Axial) 1 x 7.92 mm MG 34 in Hull, ball mount
(Optional) 1 x 7.92 mm MG 34 on Cupola ring
1 x 9 mm MP 38 or 40
1 x Nahverteidigungswaffe

Stowage

Ammunition, main armament: 82 rds
Ammunition, secondary armament: 4200 rds. 7.92 mm
Internal Fuel capacity: 730 Litres.

Armour

Type: Rolled homogeneous plate interlocking stepped joints, welded construction.
Hull, Nose upper: 80 mm at 35 degrees
Nose lower: 80 mm at 55 degrees
Sides, upper: 40 mm at 60 degrees
lower: 40 mm at 90 degrees
Rear, 40 mm at 60 degrees
Deckings: 15 mm at 0 degrees
Belly, Front: 20 mm at 0 degrees
Rear: 13 mm at 0 degrees
Turret Front: 100 mm at 80 degrees
Sides: 45 mm at 65 degrees
Rear: 45 mm at 62 degrees
Roof: 15 mm at 6 and 0 degrees
Mantlet: 120 mm

Panzerkampfwdgen Tiger Ausf B (Sd Kfz 182)

Crew: 5

Weight: Combat Loaded 69.7 tons.

Performance

Speed, Max. Road: 38 Kmph.
Cross-country: 17 Kmph.
Max. Gradient: 35 degrees.
Fording depth: 160 cm.
Trench crossing: 250 cm.
Step: 85 cm.
Min turning circle: 4.8 metres.
Range, Road: 110 km.
Cross-country: 85 km.
Power to weight ratio: 10.1 HP/ton.

Dimensions

Length overall: 1026 cm.
Length (excluding gun): 726 cm.
Width: 375 cm.
Height: 309 cm.
Ground clearance: 50 cm.

Trackwork

Length on ground: 413 cm.
Width: 80 cm.

Mechanical Details

Engine: Maybach HL 230, 23.88 litres, water-cooled, 600 bhp at 2,600 rpm.
Fuel: 860 litres.
Steering: Henschel L801 with 2 radii of turn in each gear.
Suspension: Torsion bar, resilient steel wheels.
Gearbox: Maybach-Olvar 401-216, 8 forward, 4 reverse gears.
Transmission: Pre-selector type.

Armament

Main: 1 x 8.8 cm KwK 43 L/71.
Traverse: 360 degrees, power or hand.
Secondary: 3 x 7.92 mm MG 34.

Stowage

Main ammunition: 40 rounds (APBC), 40 rounds (HE).
MG ammunition: 5,850 rounds.

Armour

Hull, nose: 100 mm at 40 degrees.
glacis: 150 mm at 40 degrees.
side: 60-80 mm at 90 degrees.
rear: 80 mm at 60 degrees.
roof: 40 mm at 0 degrees.
floor: 25-40 mm at 0 degrees.
Turret, mantlet: 80 mm.
front: 185 mm at 80 degrees.

side: 80 mm at 69 degrees.
rear: 80 mm at 70 degrees.
roof: 44 mm at 0-10 degrees.

Panzerkampfwagen Maus

Crew: 5.
Weight: combat loaded 188 tons.
Weight: Turret 50 tons.

Performance

Speed, Max. Road: 20 Kmph.
Max. Gradient: 30 degrees.
Fording depth: 200 cm.
Trench crossing: 450 cm.
Step: 72 cm.
Range Road: 190 km.
Power to weight ratio: 6.35 HP/ton.

Dimensions

Length, overall: 1008 cm.
Length (excluding gun): 903 cm.
Width: 367 cm.
Height: 366 cm.
Ground clearance: 54 cm.

Trackwork

Length on ground: 588 cm.
Width: 110 cm.

Mechanical Details

Engine: Daimler-Benz MB 509, 1200 hp at 2,400 rpm,, V-12.
Fuel: 4,800 litres.
Steering: Electric
Suspension: Independent sprung bogies.
Transmission: Two electric motors.

Armament

Main: 1 x 15 cm KwK 44 L/38 and
1 x 7. 5 cm K44 L/36.5
Secondary: 2 c 7.92 cm MG 34.

Stowage

Main ammunition: 50 rounds (15 cm), 200 round (7.5 cm) or 68 rounds if 12.8 cm gun fitted.

Armour

Hull, nose: 200 mm at 60 degrees
front: 200 mm at 35 degrees.
side: 80 + 100 mm at 90 degrees.
upper side: 180 mm at 90 degrees.
rear: 160-165 mm at 50-60 degrees.
roof: 100 mm at 0 degrees.
floor: 40 mm at 0 degrees.
Turret, mantlet: 240 mm.
sides: 200 mm at 60 degrees.
roof: 60 mm at 0 degrees.

Panzerfampfwagen E.100.

Crew: 5
Weight, Combat loaded: 140 tons.

Performance

Speed, (Prototype) Max. Road: 23 Kmph.
(Series) Max. Road: 40 Kmph.
Fording depth: 190 cm.
Step: 90 cm.
Min. Turning circle. 8 metres.
Ground pressure 1.4 Kg/sq.cm.
Range (internal fuel) Road: 120 Km.
Power to weight ratio: (Prototype) 5 metric HP/ton.
(Series) 8.5 metric HP/ton.

Dimensions

Length overall: 1027 cm.
Length: 870 cm.
Width overall (Battle order) 448 cm.
Width: 329 cm.
Height: 332 cm.
Ground clearance: 50 cm.
Fire height of gun: 254 cm.
Turret ring dia: 306 cm.

Trackwork

Centres: (Battle Tracks) 307.5 cm.
Centres: 292 cm.
Length on ground: 490 cm.
Width: (Battle Tracks) 100 cm.
Width: 5.5 cm.

Mechanical Details

Engine: (In Prototype) 1 x Maybach HL 230 P30 V 12 (60 degrees) OHV Water cooled Petrol of 23.88 litres developing 700 bhp (metric) at 3000 rpm.
(In Series) 1 x Maybach HL 234 developing 1200 bhp at 300 rpm.
Transmission: (In Prototype) Maybach Olvar type B40 12 16 8F - 4R speeds.
(In Series) Maybach Mekydro 8F - 4R speeds.

Armament

Mains: 1 x 17 cm KwK 44 or (15 cm in DRG)
Calibre, and length in calibres: 174/150 mm (6.85/5.9 ins. L/38.
Traverse: 360 degrees turret.
Elevation: plus 23 degrees, minus 7 degrees.
Secondary armament: 1 x 7 cm KwK 44 L/36.5 Co-Axial.
1 x 7.92 mm MG 34 Co-Axial.
2 x 9 m.p. 44- Ball mounts in turret sides.

Stowage

Ammunition: Main Armament: 32 rds in turret - total unknown.
Ammunition: 7.5 cm 200 rds.
Internal Fuel capacity 1200 Litres.

Armour

Type: Rolled homogeneous plate interlocking stepped joints, welded construction.
Hull, Nose upper: 200 mm at 30 degrees.
Nose lower: 50 mm at 50 degrees.
Sides, upper: 120 mm at 90 degrees
Rear: 150 mm at 60 degrees
Decking: 40 mm at 0 degrees.
Belly, Front: 80 mm at 0 degrees.
Rear: 40 mm at 0 degrees.
Turret Front: 240 cm Rounded
Sides: 200 cm at 60 degrees.
Rear: 200 mm at 83 degrees.
Roof: 40 mm at 0 degrees.
Mantel: 200 mm Average.

Scale drawings

1:76 scale (4mm = 1ft)

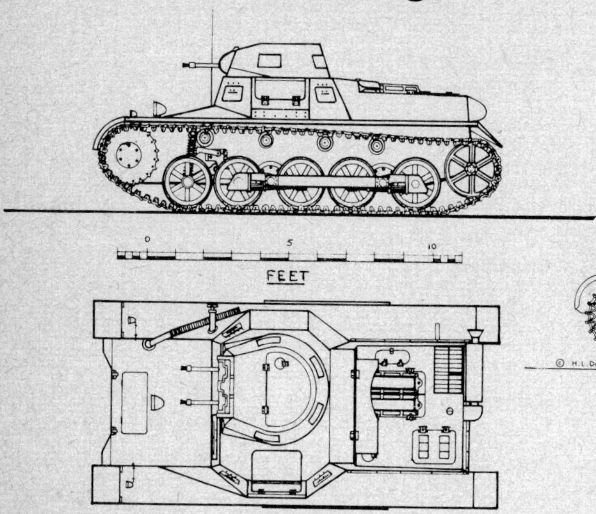

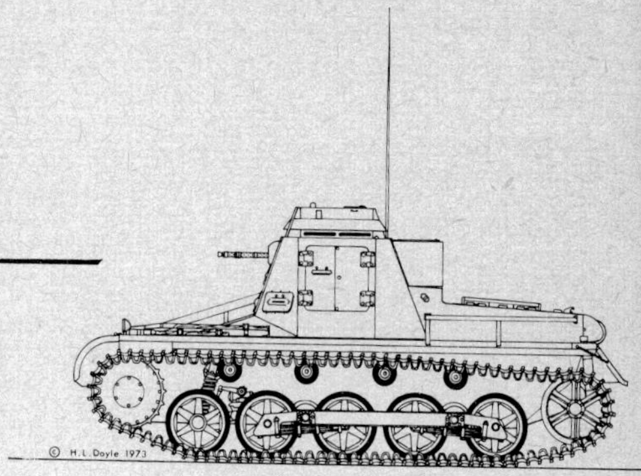

Kl PzBefWg I Ausf B

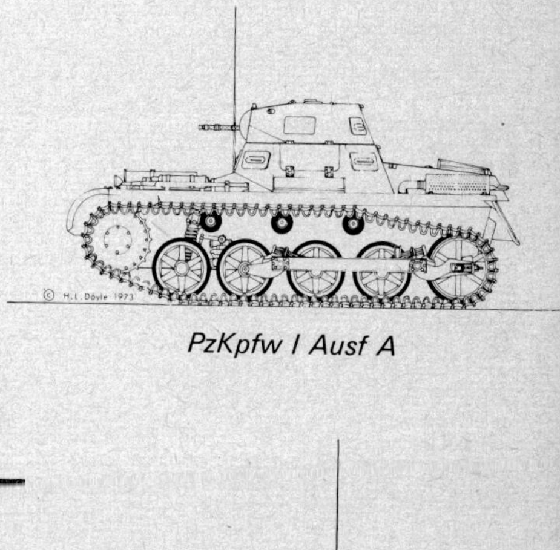

PzKpfw I Ausf A

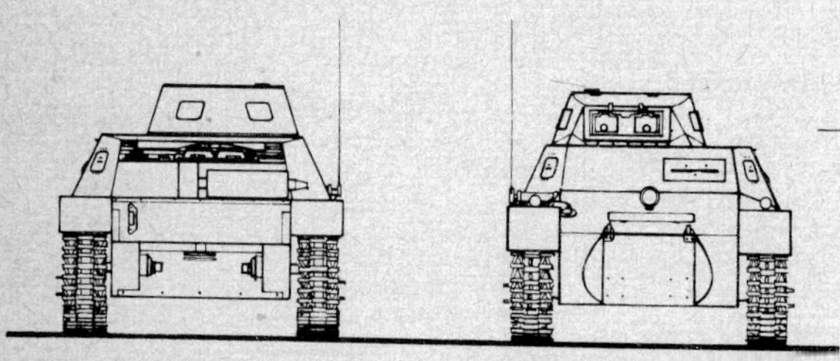

PzKpfw I Ausf B

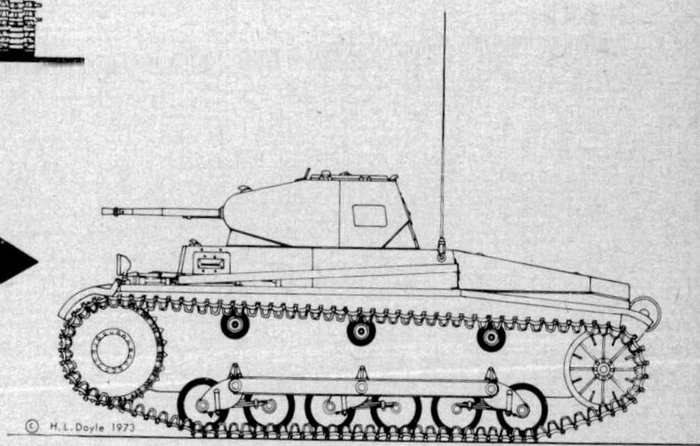

PzKpfw II Ausf b ▶

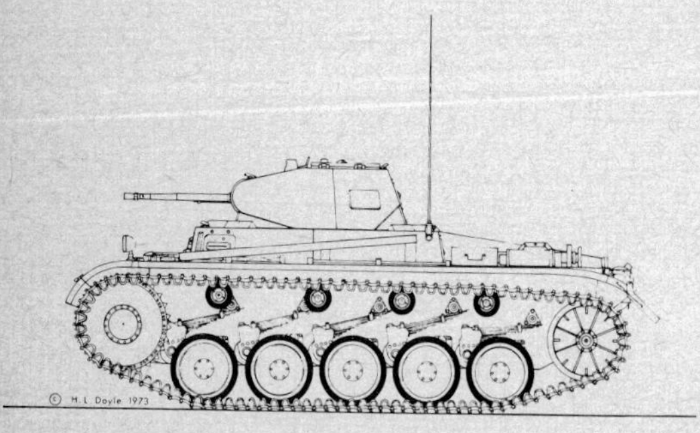

◀ *PzKpfw II Ausf A, B, C*

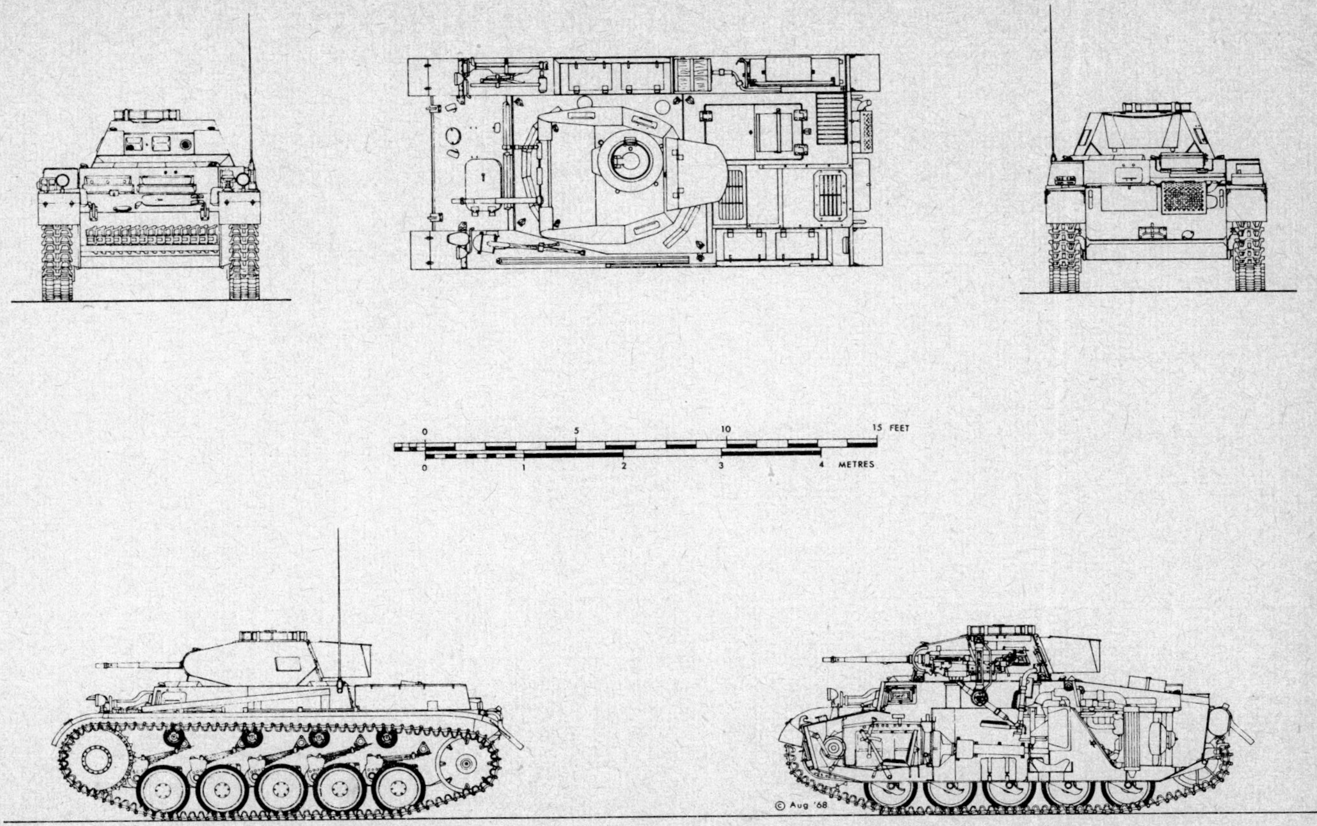

PANZERKAMPFWAGEN II AUSF. F.

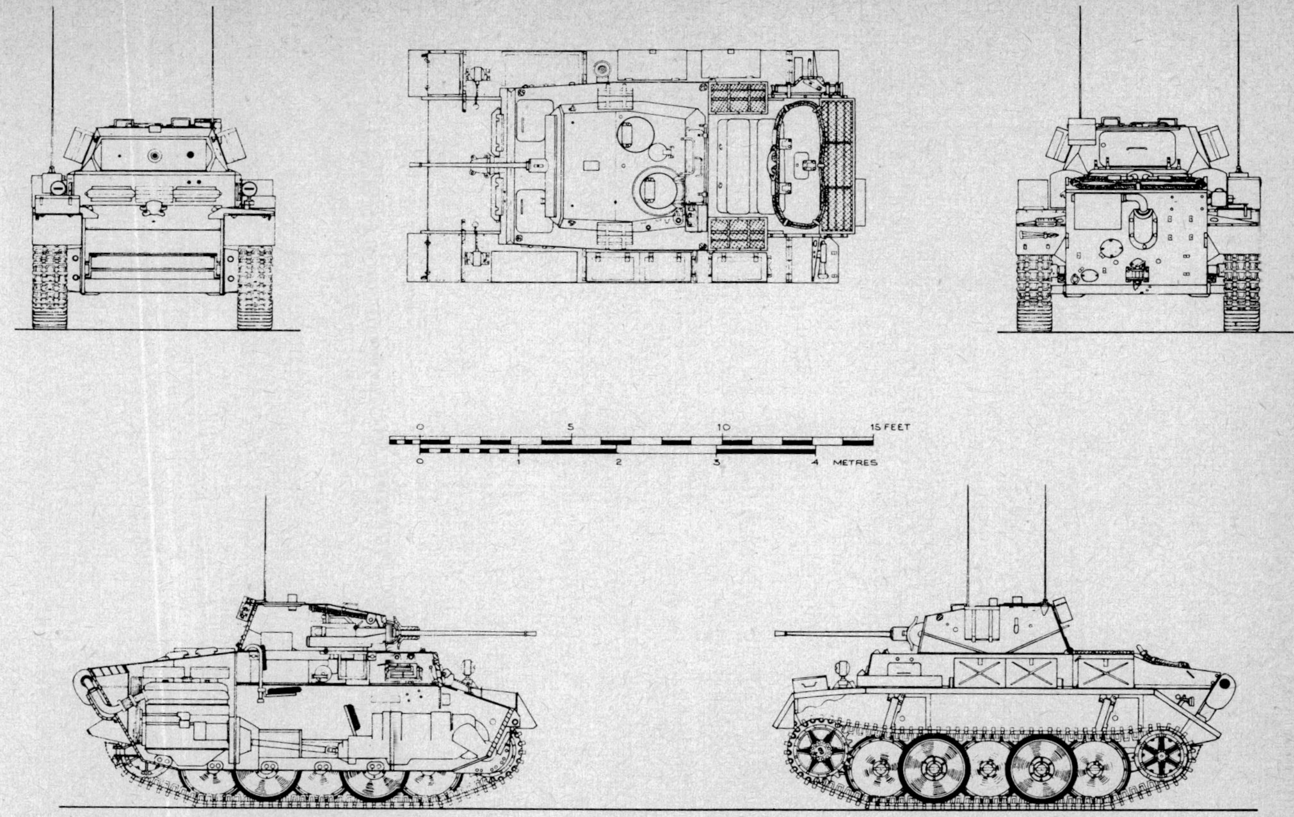

PANZERSPÄHWAGEN II (2cm. KwK 38) 'LUCHS'

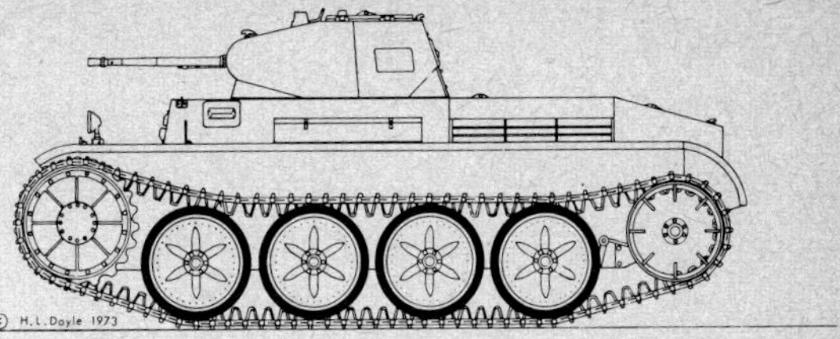

PzKpfw II Ausf D, E ▶

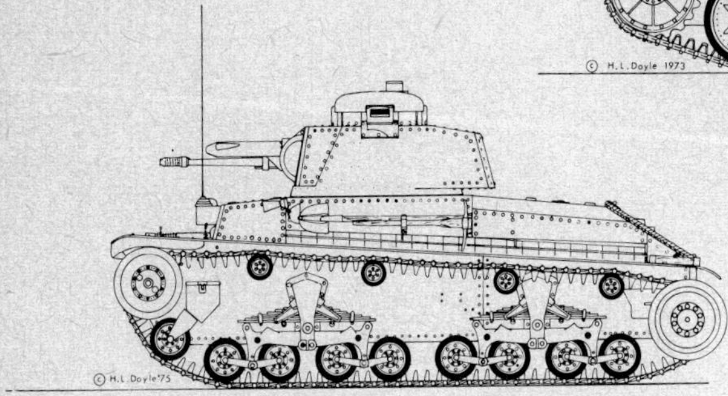

◀ *PzKpfw 35(t)*

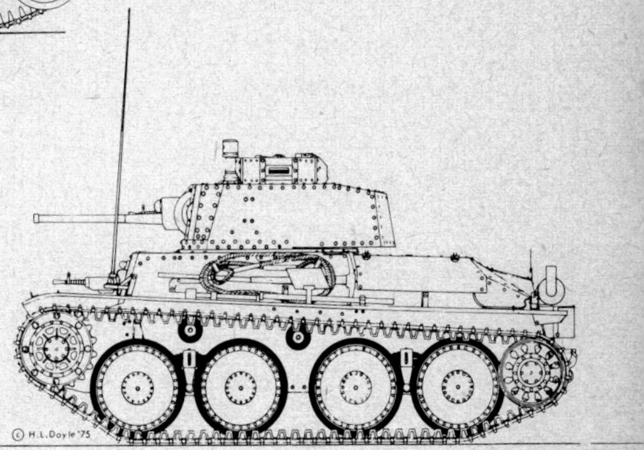

PzKpfw 38(t) Ausf G ▶

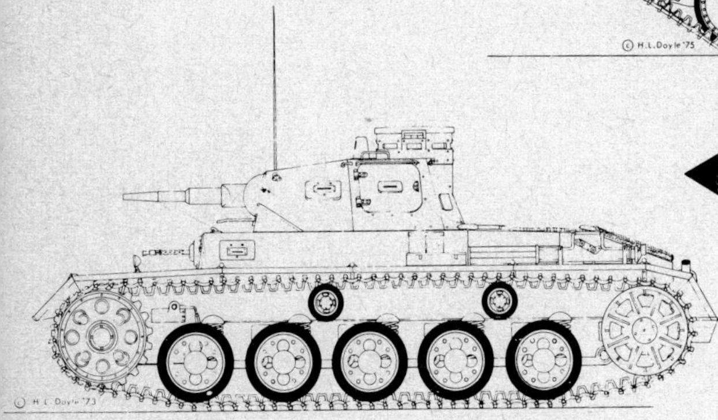

◀ *PzKpfw III Ausf A*

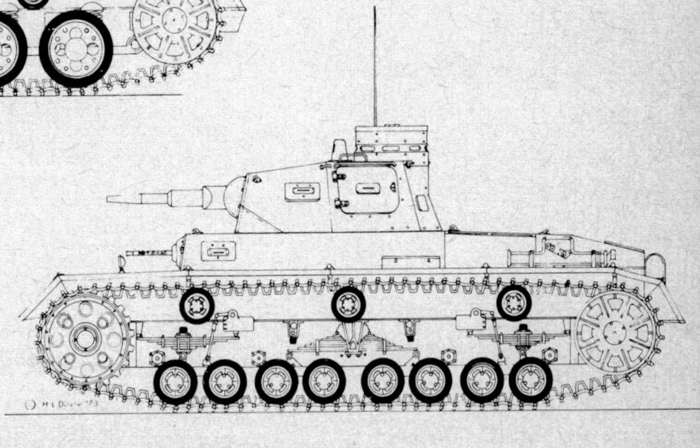

PzKpfw III Ausf C ▶

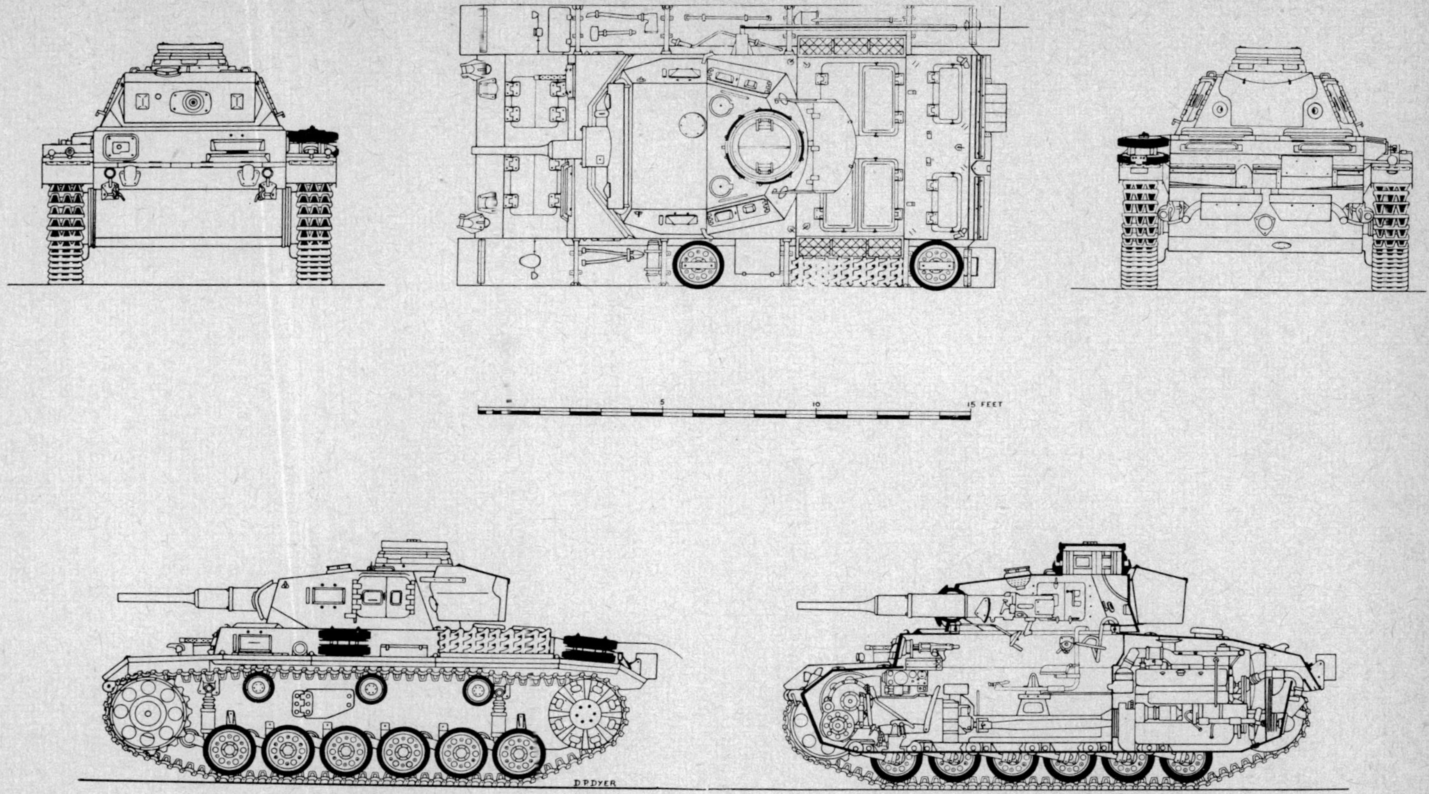

PANZERKAMPFWAGEN III (5 cm) AUSF F

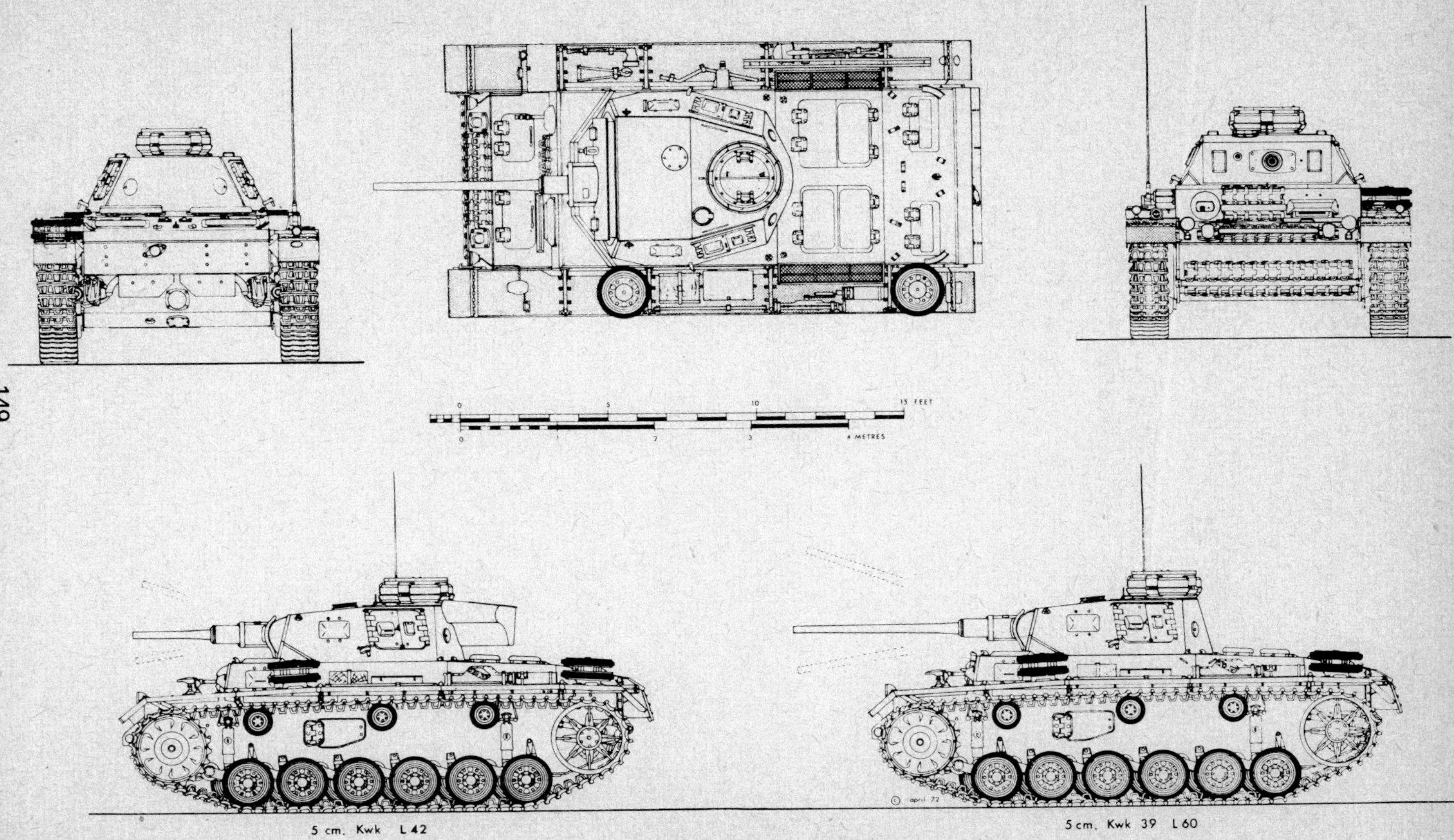

PANZERKAMPFWAGEN III AUSF. J.

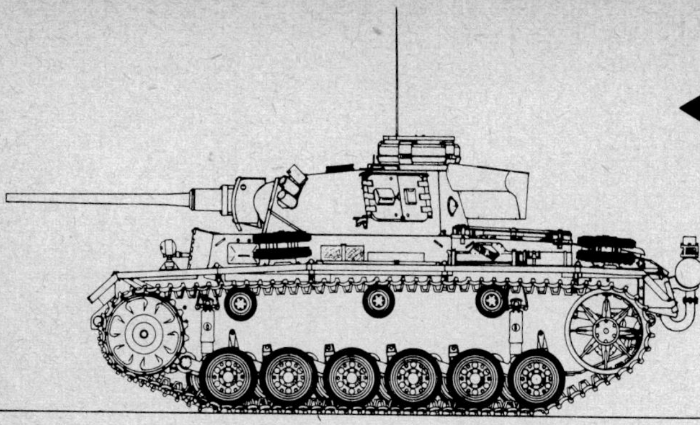

PzKpfw III Ausf L

PzKpfw III Ausf N

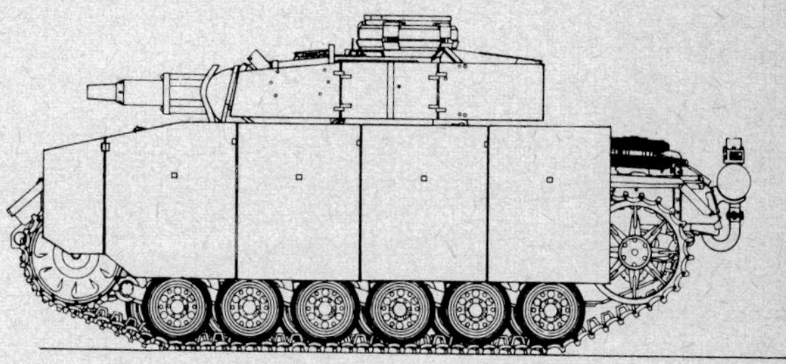

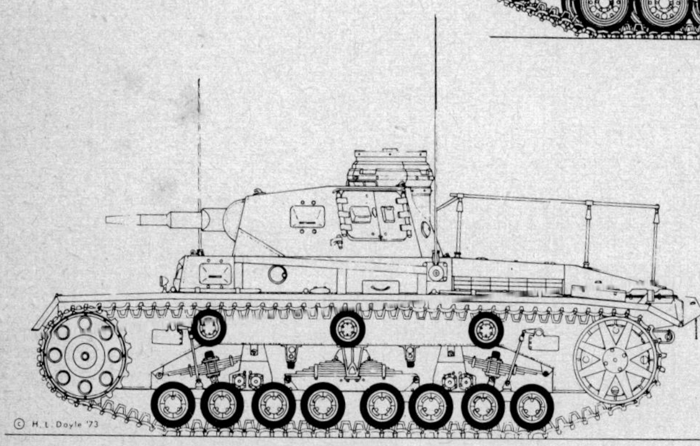

PzBefWg III Ausf D1

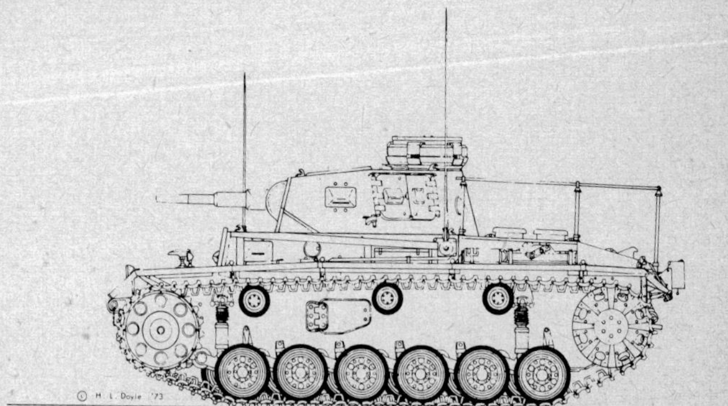

PzBefWg III Ausf H

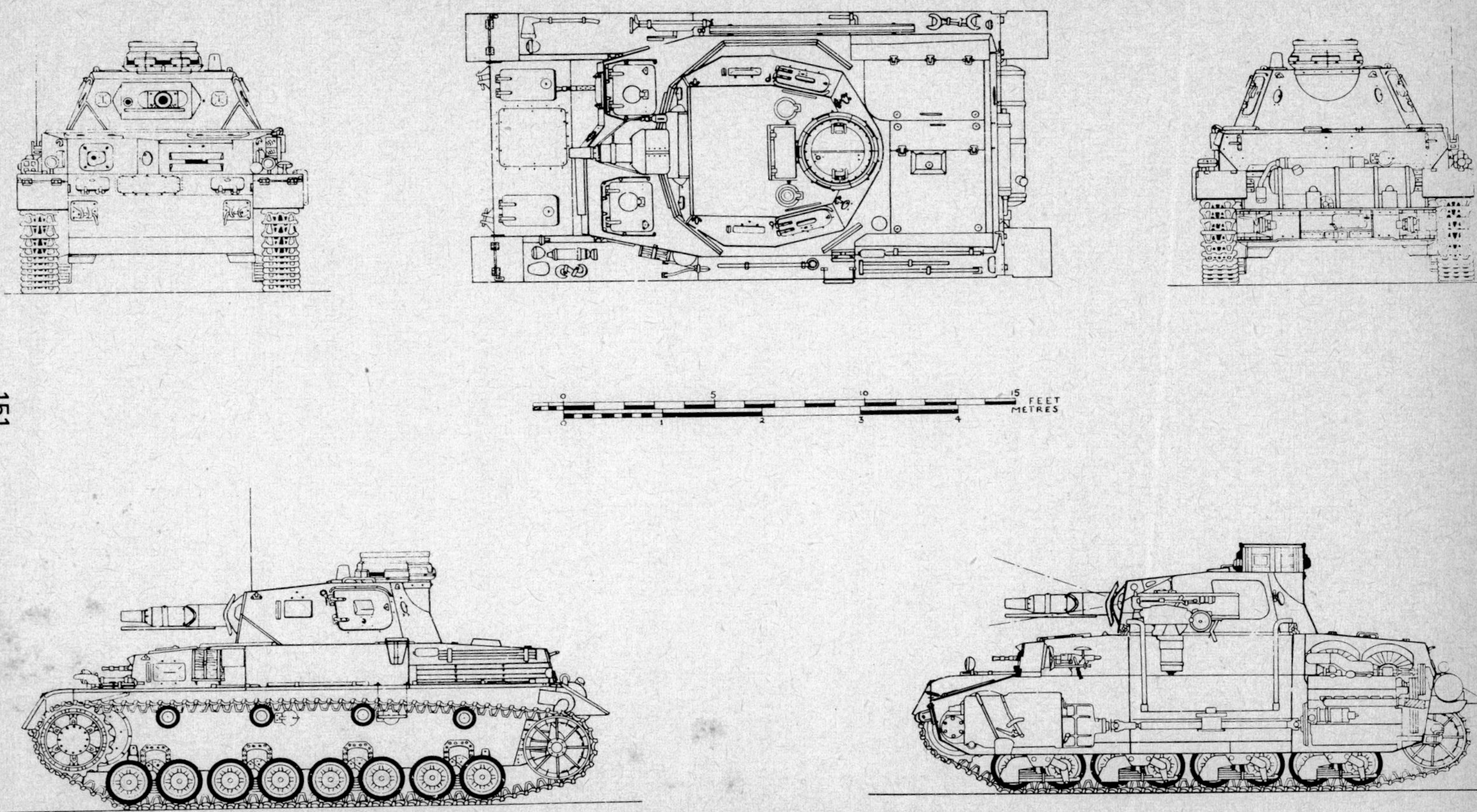

PANZERKAMPFWAGEN IV AUSF D

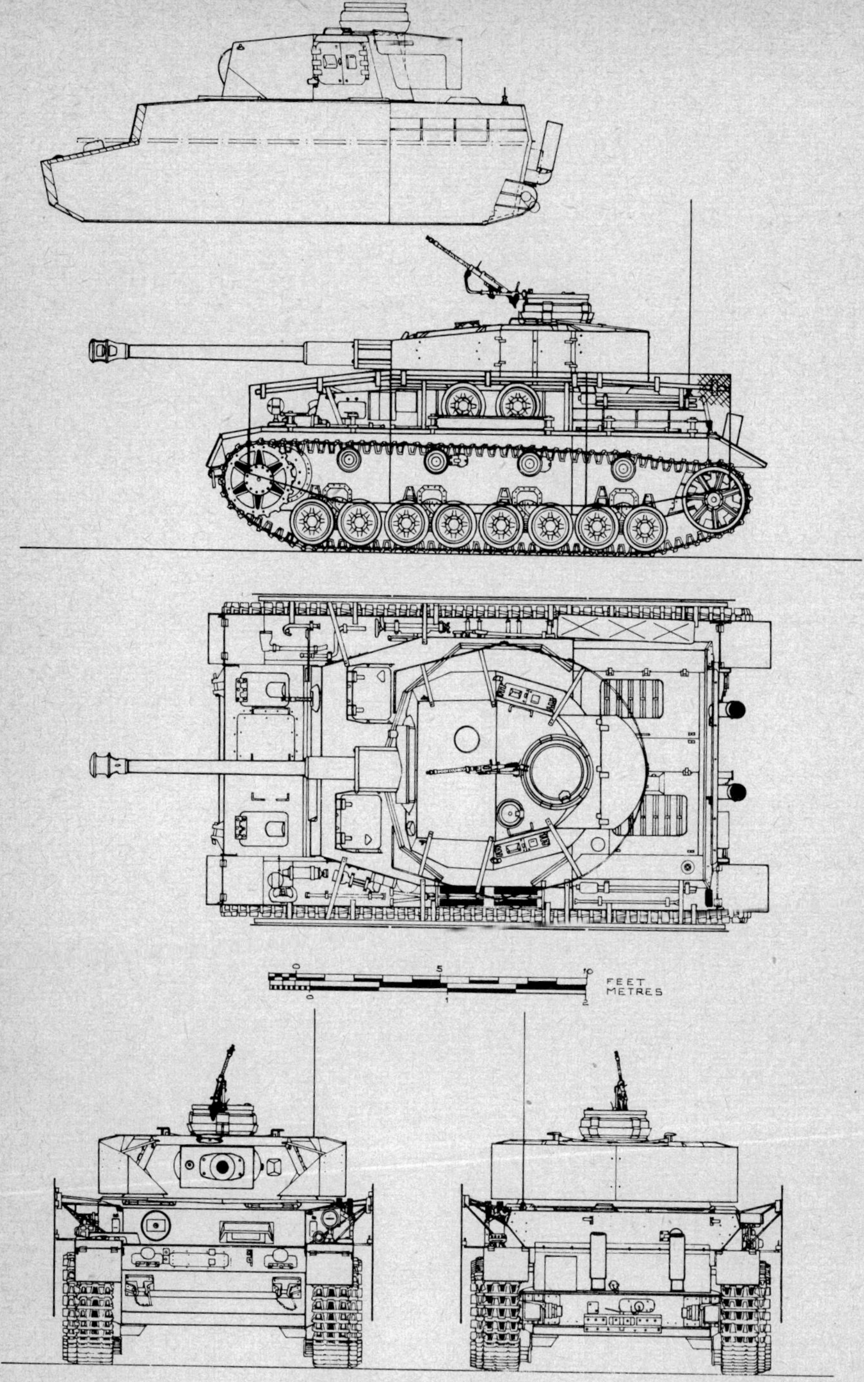

PANZERKAMPFWAGEN IV AUSF J

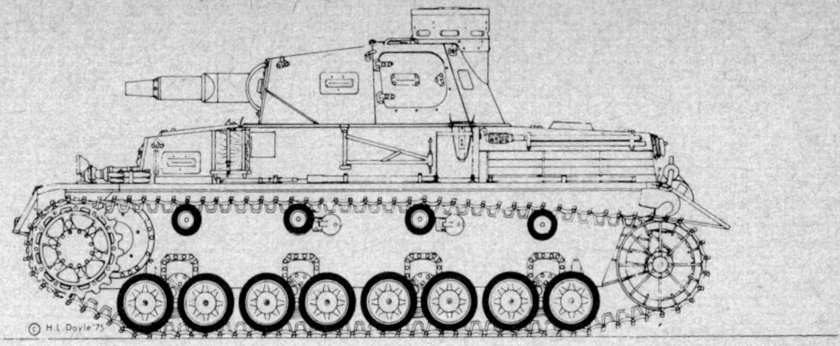

PzKpfw IV Ausf A

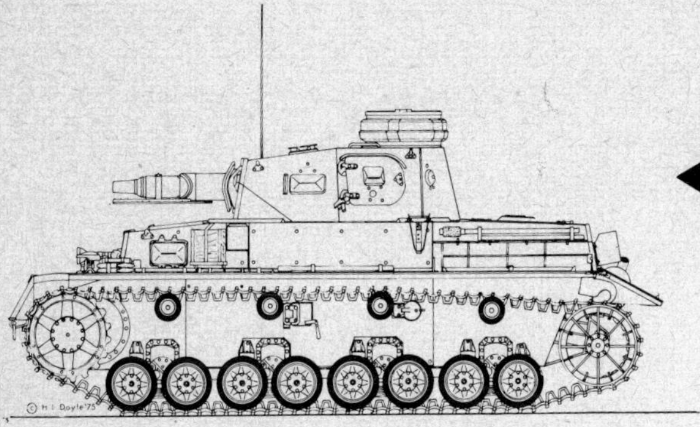

PzKpfw IV Ausf E

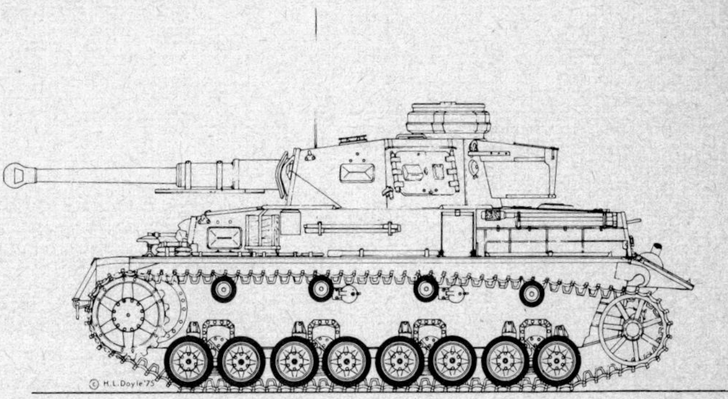

PzKpfw IV Ausf F2

PzKpfw IV Ausf H (early)

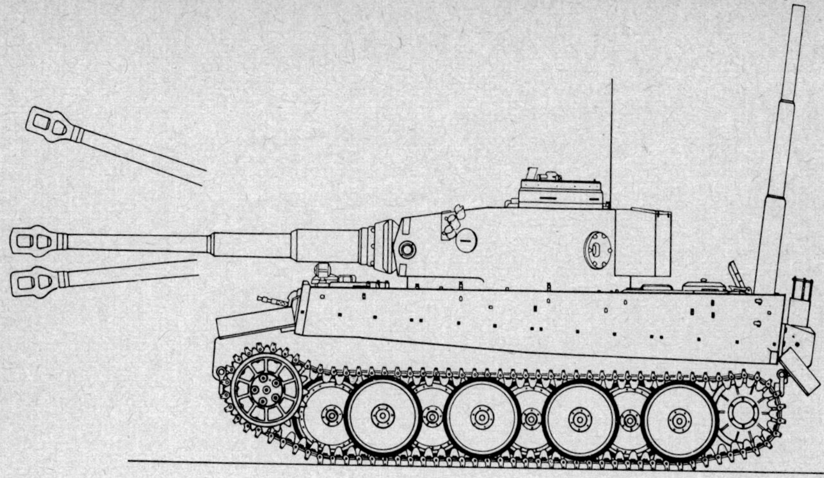

LEFT HAND SIDE TRANSPORTATION TRACKS, SNORKNEL RAISED, AIR PRECLEANERS AND FLASH GUARDS REMOVED.

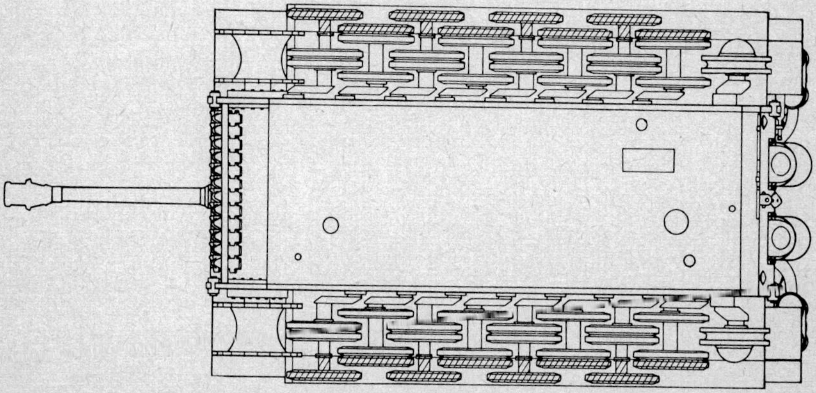

PLAN VIEW FROM BELOW TRACKS NOT FITTED. SHADED WHEELS REMOVED WHEN TRANSPORT TRACKS USED

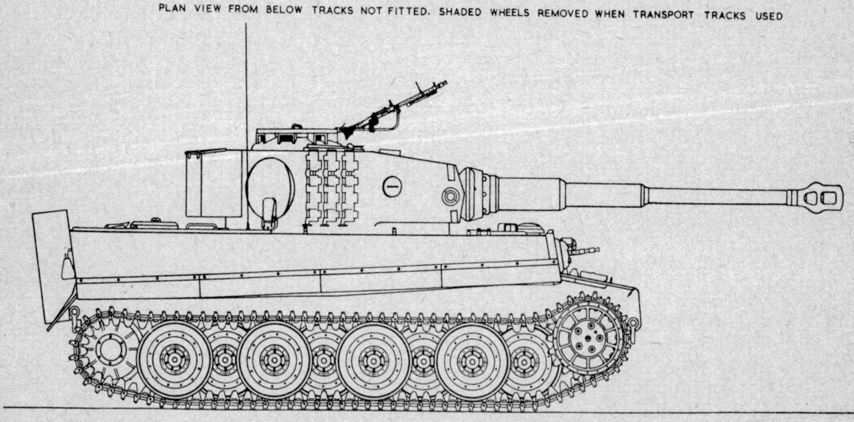

FINAL VERSION RIGHT HAND SIDE

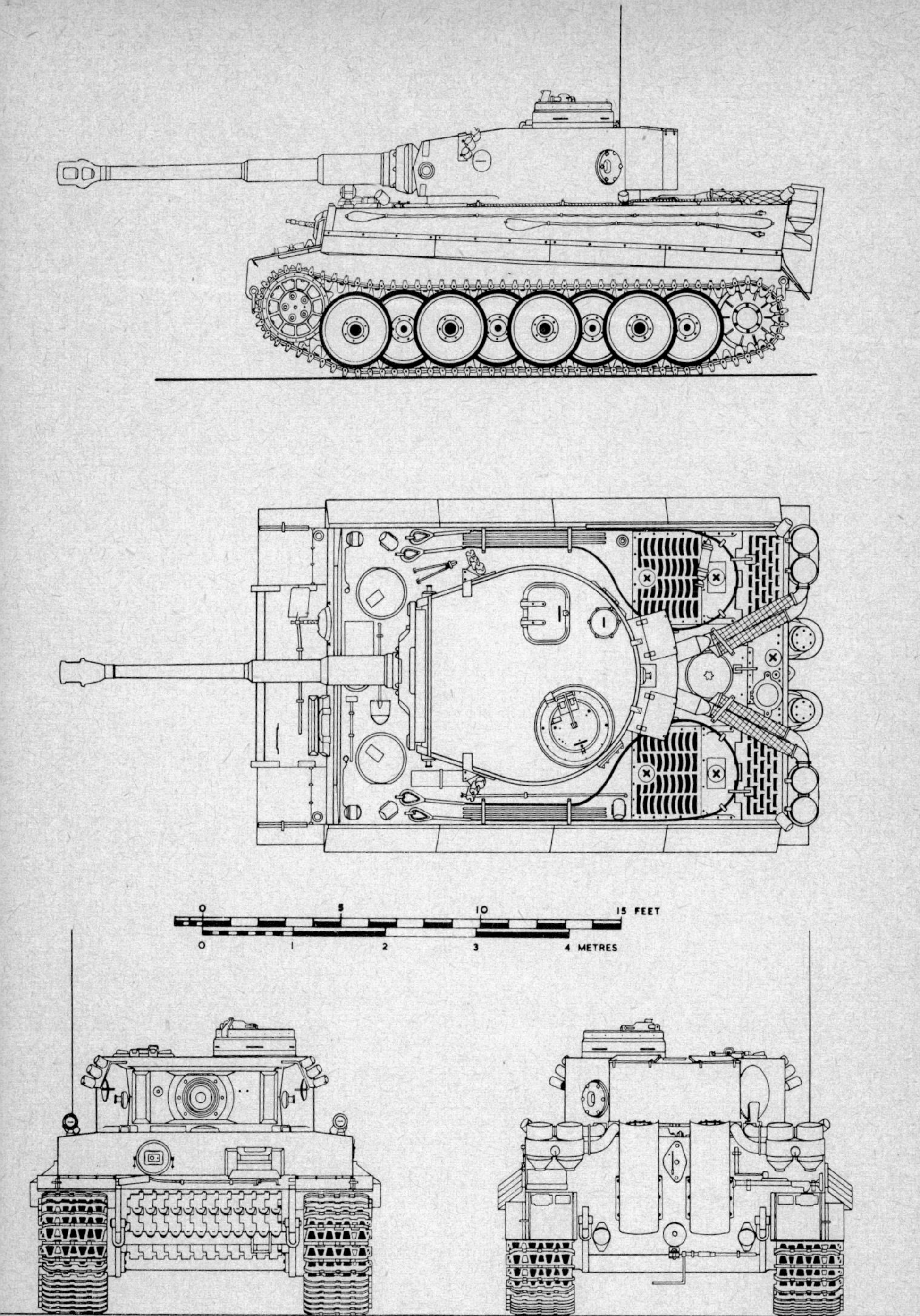

PANZERKAMPFWAGEN IV, 'TIGER' AUSF E

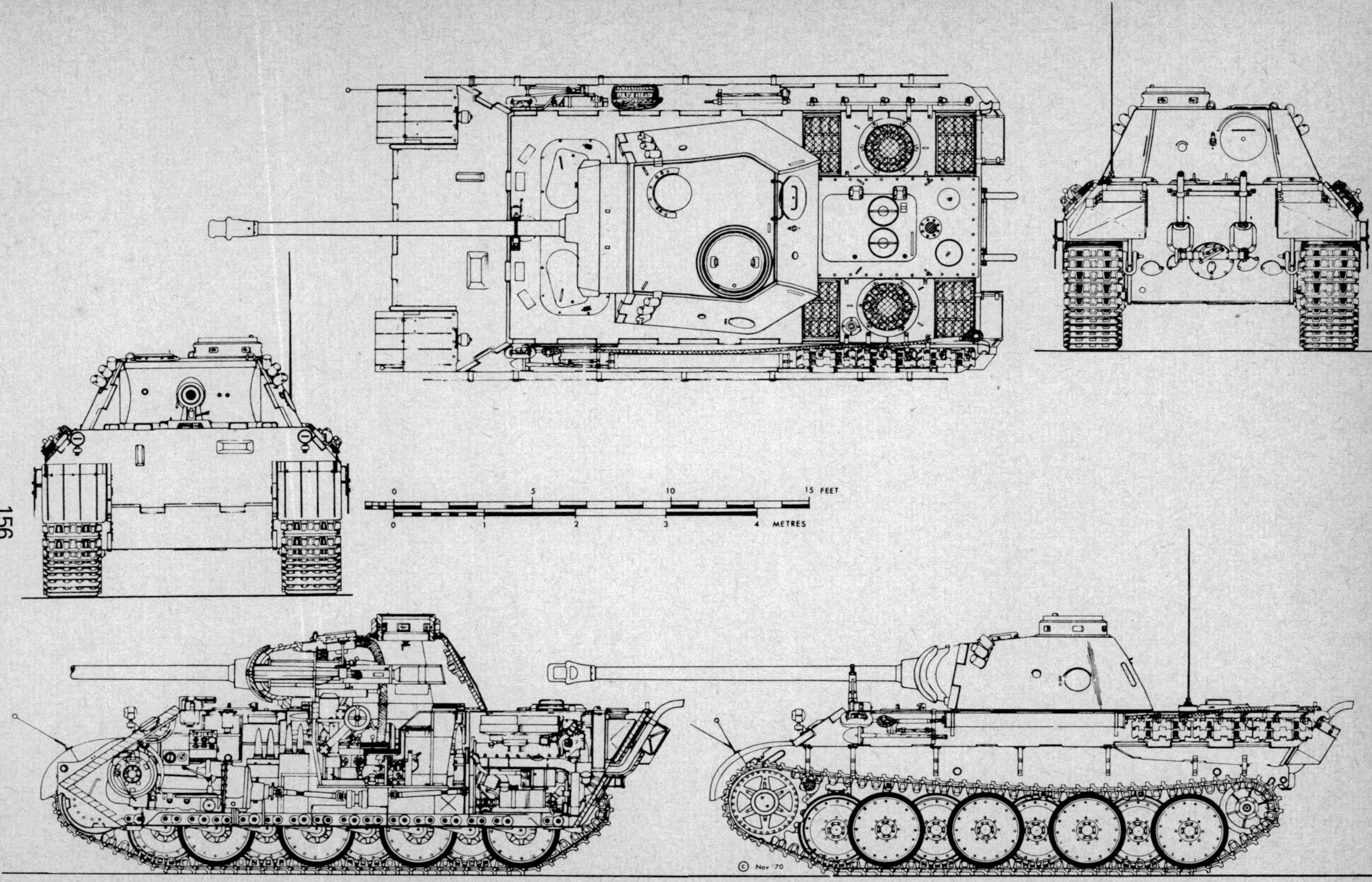

PANZERKAMPFWAGEN 'PANTHER' AUSF. D

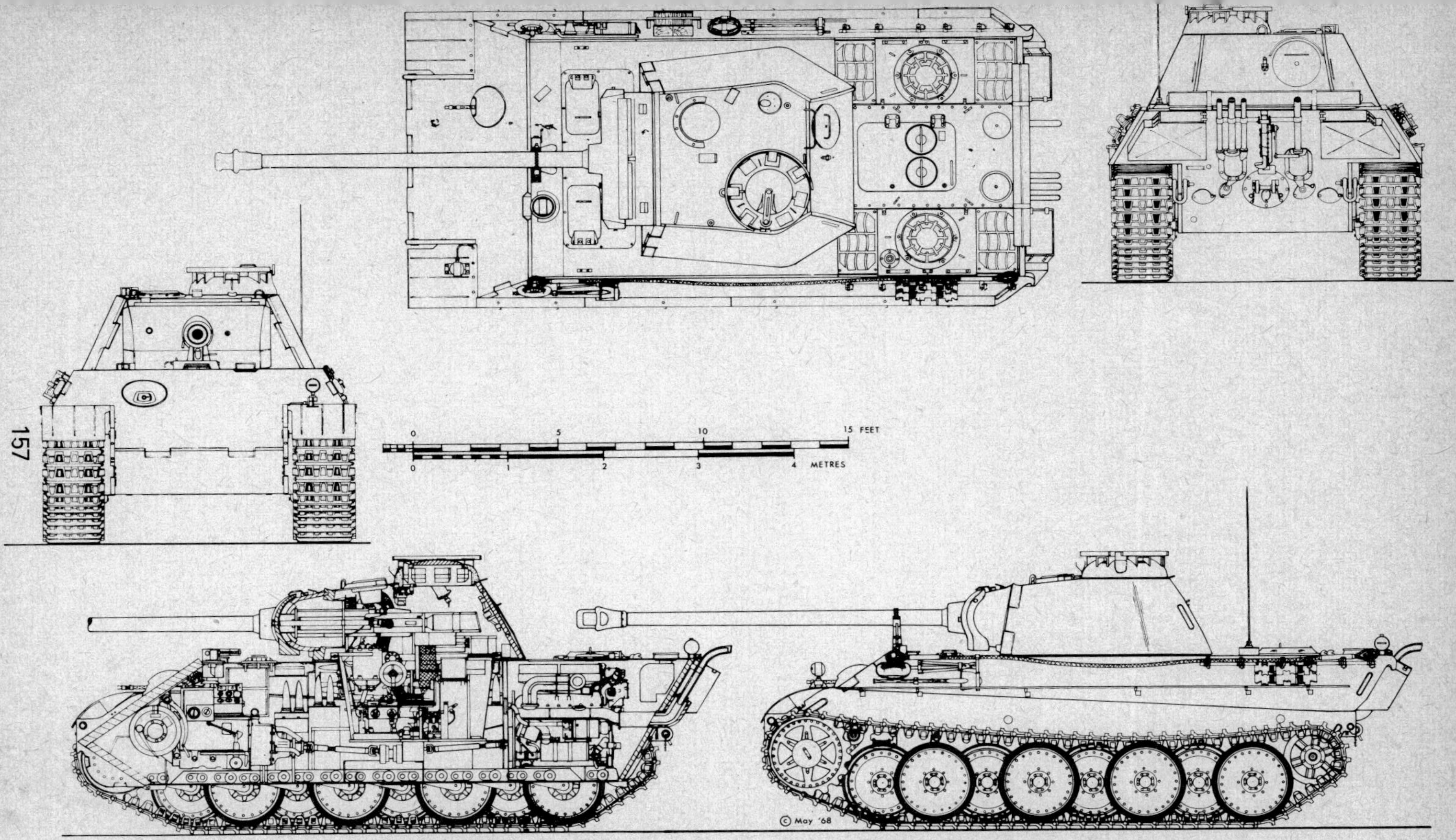

PANZERKAMPFWAGEN 'PANTHER' AUSF. G

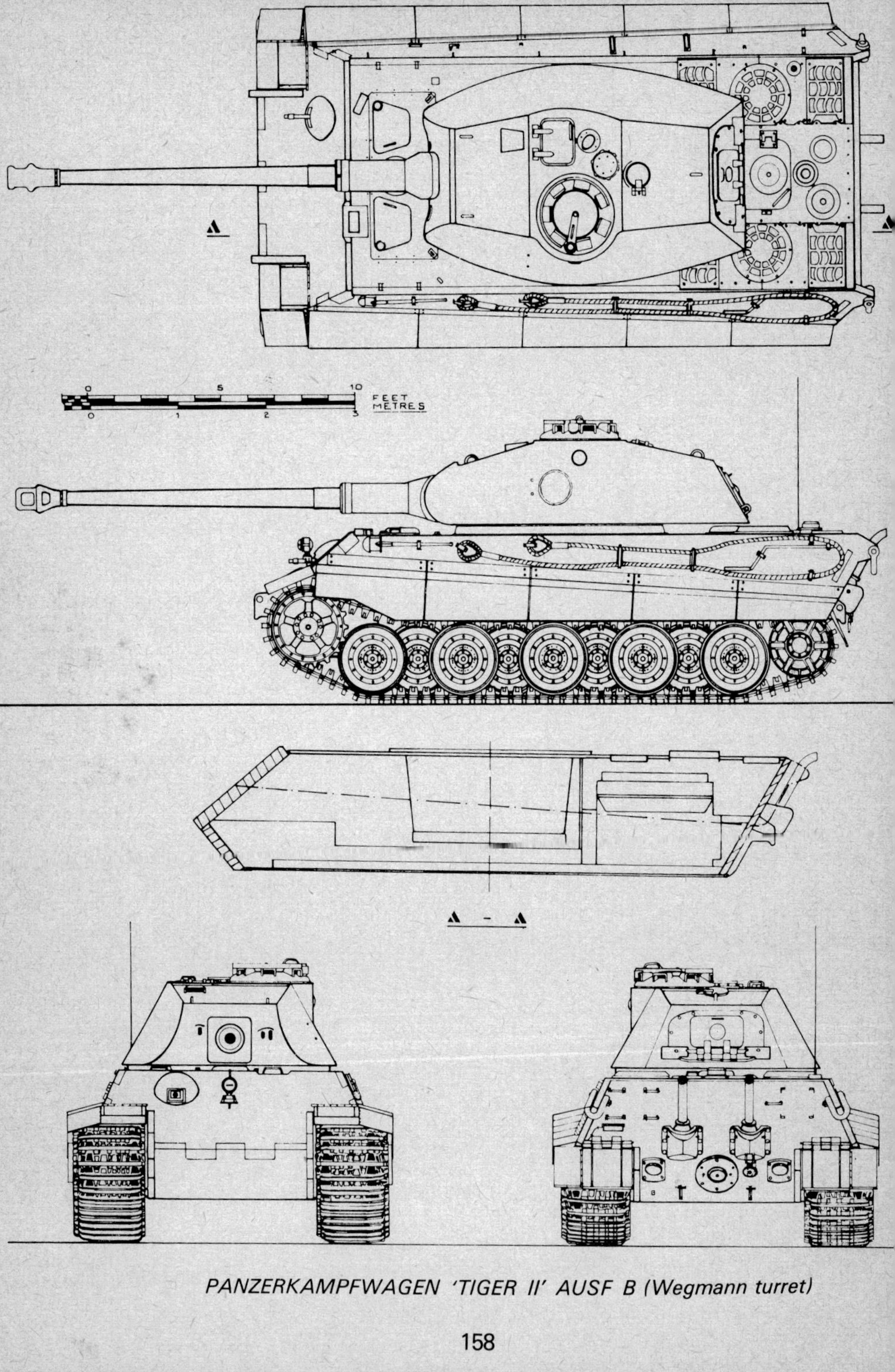

PANZERKAMPFWAGEN 'TIGER II' AUSF B (Wegmann turret)

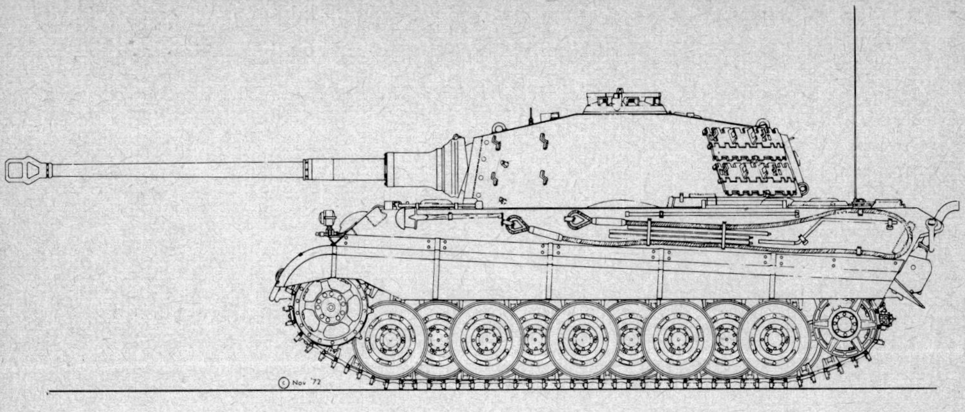

PzKpfw Tiger II Ausf B (Henschel turret)

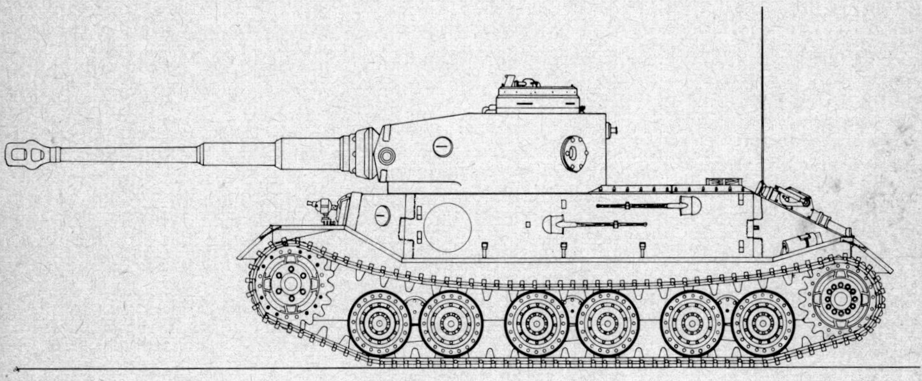

VK 4501 (P), Tiger (P)

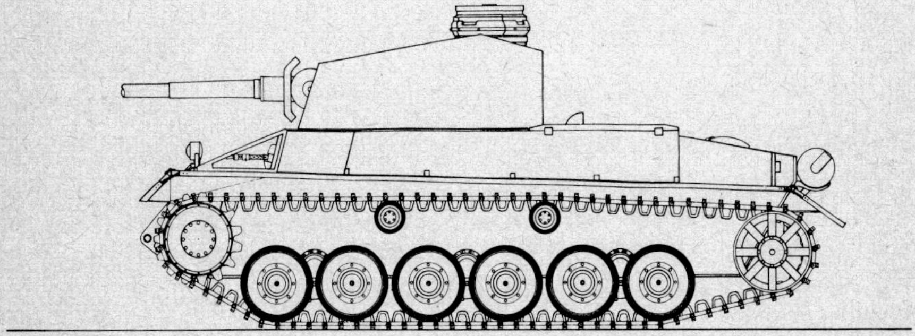

VK 3001 (P), Porsche Typ 100, Leopard

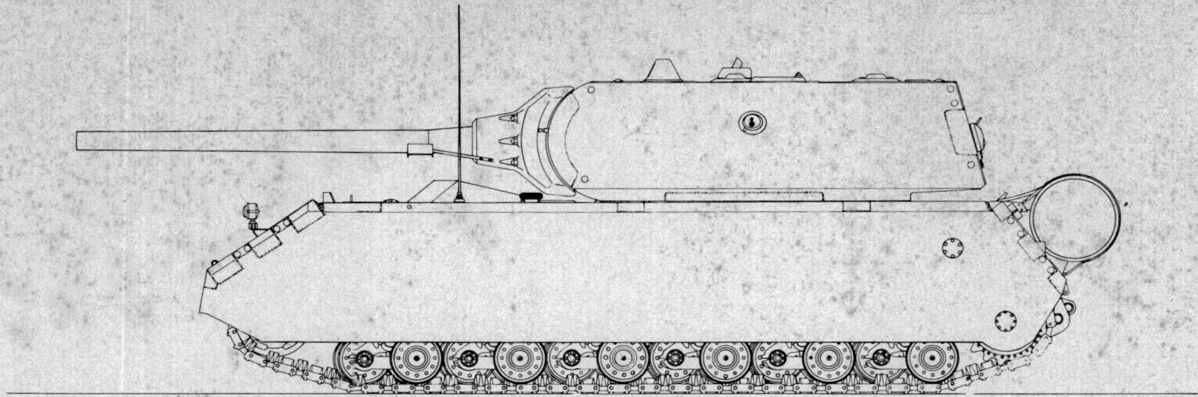

PzKpfw 'Maus'

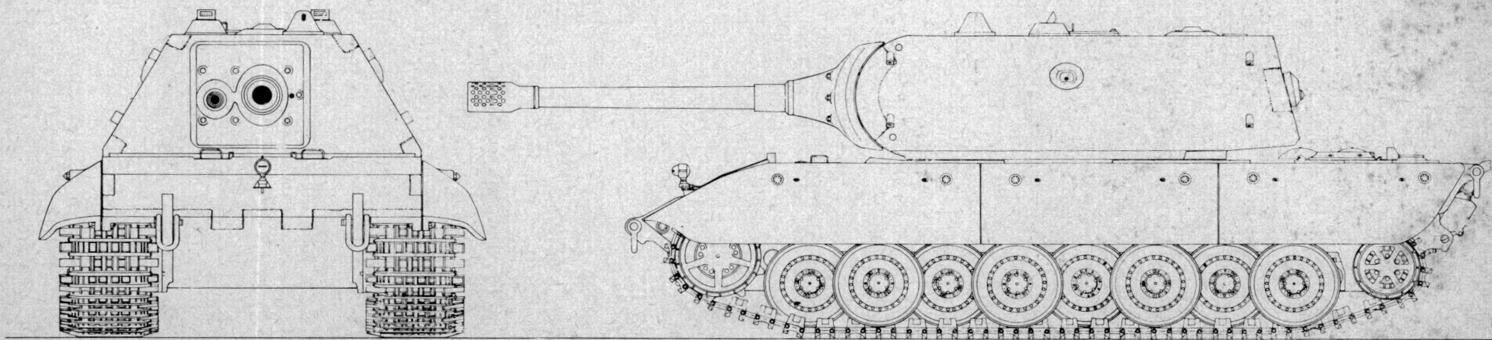

PzKpfw E-100